Also by Angela Bonavoglia
The Choices We Made:
Twenty-Five Women and Men Speak Out About Abortion

GOOD CATHOLIC GIRLS

HOW WOMEN ARE LEADING THE FIGHT TO CHANGE THE CHURCH

ANGELA BONAVOGLIA

ReganBooks
An Imprint of HarperCollins*Publishers*

A hardcover edition of this book was published in 2005 by ReganBooks, an imprint of HarperCollins Publishers.

FIRST PAPERBACK EDITION PUBLISHED 2006.

Designed by Erin Benach

The Library of Congress has catalogued the hardcover edition as follows:
Bonavoglia, Angela.
  Good Catholic girls : how women are leading the fight to change the church / Angela Bonavoglia.—1st ed.
    p. cm.
  Includes bibliographical references (p.) and index.
  ISBN 0-06-057061-X
  1. Church renewal—Catholic Church. 2. Change—Religious aspects—Catholic Church. 3. Catholic women. I. Title.

BX1746.B66 2005
282'.082—dc22

2004064936

ISBN 13: 978-0-06-057063-7 (pbk.)
ISBN 10: 0-06-057063-6 (pbk.)

06 07 08 09 10 RRD 10 9 8 7 6 5 4 3 2 1

CONTENTS

PREFACE TO THE
PAPERBACK EDITION

✝ — ✝ — ✝

Shortly after *Good Catholic Girls* was released in March 2005, Pope John Paul II fell gravely ill. Day and night, the faithful held vigil in St. Peter's Square. When John Paul II passed on, millions paid their respects.

Immediately, speculation abounded as to who would succeed the beloved but extremely conservative Pope. Among many of the progressive Catholic women who are the substance and soul of this book, there was great hope, but also great fear. The fear centered on one man particularly: Cardinal Joseph Ratzinger, the ultra-conservative prefect of the Vatican Congregation for the Doctrine of Faith. Ratzinger appears frequently on these pages—opposing, denouncing, even excommunicating Catholic women whose Vatican II vision of the future Church failed to match his own. On April 19, 2005, the fears of many women reformers were realized: Cardinal Ratzinger became Pope Benedict XVI.

Since this book was published, I have heard from many Catholic women. Reading about the women here and eyeing the appended list of nearly forty progressive organizations, one woman never before active in Church reform wrote: "I have decided to find my voice and use it in the Church." Another wrote that she had never intended to read the book—she bought it because someone she knew was in it. She described herself as a "late bloomer" in terms of feminism, as well as a later convert to Catholicism. There were things about the Church she couldn't buy, but she resolved to think about those "some other day." She did read the book, however, and found that day had come. Wrote another woman, after pondering the sins of the Church fathers described here, "I am not the terrible, divorced woman the Church would have you believe. I raised my children to be responsible adults and worked hard to make my life whole again. . . . We are all women of dignity and honor."

I have heard from women about their hurt and disappointment in the Church, but also about their continued love for the Church. Mostly I have heard this: I no longer feel alone.

The women in this book move me as much today as when I first met them, when these words began to form in my mind: *Blessed is she who comes in the name of the Lord.*

Angela Bonavoglia
October 2005

INTRODUCTION

✝ ⚜ ✝

*"We shall honor and support our women leaders
and prophets that their days be long in the land and
their voice be heard among the people."*
Elisabeth Schüssler Fiorenza
"Ten Statements of Feminist Commitment"
Women Moving Church Conference, 1981

I grew up Catholic. As a girl, I spent a lot of time, perhaps an inordinate amount of time, visiting the basement chapel of Saint Ann's Church, atop the sprawling monastery grounds in Scranton, Pennsylvania. I can still hear the click of my shoes on the marble floor. Feel the cool air against my face. I'd settle alone in the near darkness, before the long, sleek body of Jesus Christ—crucified on one side of the altar, risen on the other. Resting my child's knees on the soft leather kneeler, leaning my hands on the wooden pew, I'd chat with God. I told God everything. And God listened. I've taken those moments of trust and peace with me all my life.

But that was my private Catholic Church. As I grew up, I saw how miserably the public Church failed to live up to its own ideals, deeply instilled in me, of justice and equality. I came to deplore the Church's demonization of sexuality, its arrogance, and its hypocrisy. I saw the Church as depriving women of authority not only in the public sphere, by forbidding women's ordination and access to the highest levels of sacramental and jurisdictional power but also in the private sphere, by usurping a woman's right to her own conscience and moral voice on matters sexual, marital, and maternal.

By the time I was in my twenties, I became a Catholic in revolt. Later, I marched. I wrote. But somewhere along the way, I gave up. I became a Catholic in exile, and a Catholic trapped. Being Catholic isn't something I can change, any more than I can change the configuration of my bones. I love the sensuality of the religion—the imagery, the incense, the music, and the Mass. I love the saints, especially Mary's mother, Ann, to whose grotto I wore a path. I find enormous comfort in the Resurrection, and could not live without the Eucharist. It is food for my journey.

Today, I am an itinerant Catholic. I make my way to Mass as often as I can, hoping the sermon will not drive me away. I go on Christmas and Easter because they celebrate beginnings and hope; on Good Friday because I need help facing death; and on many other days, when I am especially grateful, when I am lost, and when I have been brought to my knees by life. I pray everywhere and all the time. I have long known that born a Catholic, I will die a Catholic. I had been resigned to dying with a broken heart.

Then, in 2001, I thought I detected a shift in the winds. I heard about a woman—a nun, no less—who found the courage to say no to the pope. I was shocked to think that change might be possible. That woman, Sister Joan Chittister, was the reason I arrived at the Adam's Mark Hotel in Philadelphia on a Saturday morning in September 2001. The twenty-fifth-anniversary conference of the then largest Roman Catholic reform organization Call to Action was in full swing. Just four days had passed since the attacks on the World Trade Center. Despite the catastrophe that had befallen America, 1,100 people turned out. While my interest in one woman working for change in the Church brought me to that conference, I learned that there were many such women, all over the world.

Principled, brave, and brimming with well-earned indignation, they have inspired a movement, a literature, and a network of organizations hell-bent on restoring women and lay Catholics to their rightful place in the Church. Devoted nuns, newly ordained women priests, lay ministers, theologians, and activists, they are passionate advocates all. Like their Muslim, Jewish, and Christian sisters, they have stepped forward to challenge one of the last and most impenetrable bastions of male authority.

I began to follow these women and the global movement they represent at a time of unprecedented upheaval in the U.S. Catholic Church, amid breaking news of an epidemic of priest child sex abuse, decades-long and nationwide. I found these women involved in all aspects of progressive Church reform, beginning with the Second Vatican Council and continuing to the present portentous moment. I found them heading up parishes; leading the fight for women's ordination; creating a new Catholic theology and reenvisioning God. They were resisting Vatican silencing, challenging the Church's sexual repression, and rescuing female Church leaders from the trash heap of history. They were bringing the Church's human rights commitment home—to gays and lesbians, to the divorced and remarried, and to the survivors of priest child sexual abuse. They were organizing Catholics never before active to call the Church to accountability and change.

In reality, the heyday of the Catholic women-led Church reform movement, in sheer numbers, occurred in the 1970s and 1980s, paralleling the rise of this nation's second wave of secular feminism. It was a time when Catholic women's organizations mounted huge conferences, to which Catholic women full of energy and hope flocked by the thousands. But the reign of Pope John Paul II, with his vision of a deeply authoritarian Church without an opportunity for the full participation of women, squelched their enthusiasm and shattered their hopes. In response, Catholic women in droves left the Church as well as the women-led Church reform groups. And while nuns led the women's movement for change in the Church originally, their numbers have declined precipitously, and they are not being replaced.

Yet, what becomes of women in the Catholic Church will dictate what becomes of the Church itself. Women make up more than half of the Church's one billion members. When they leave, so do their children. Worldwide, the Church is in crisis. The priest shortage is pressing in the United States, a major factor in parish closings but downright dire in many poor regions of the world. Catholics in large parts of Africa and Latin America have been without regular Eucharist for years. Yet, America is raiding priests from the Third World, assigning them to parishes in the United States where they are unfamiliar with the culture. Priest sex abuse scandals have shaken the Church not only in the United States but in other countries as well, such as Austria and Australia, Ireland and Canada, the Philippines and South Africa. Finances are shrinking, attendance at Mass is declining, and the Church's credibility is in tatters.

In an effort to regain its lost moral voice, the Vatican is attempting to reestablish its authority over the Catholic family. But it cannot exercise authority over the Catholic family without the cooperation of women. Hence, the Vatican in 2004 issued a startling letter to the world's bishops "On the Collaboration of Men and Women in the Church and in the World." It approved of married women having careers, as long as they maintain complete openness to motherhood and do not seek "liberation from biological determinism." It denounced women who talk about their "subordination" because it causes "antagonism"; who seek power because it disadvantages men and is "lethal" to the family; who dare to be for themselves instead of following their true "feminine" natures "to live for the other and because of the other"; and who give insufficient importance to Jesus' maleness—which, not coincidentally, is the Church's foundation for an all-male priesthood. The document makes crystal clear that controlling women—their understanding of their nature, their sexuality, and their place in the family and in the Church—is the keystone to the male clerical hierarchy's ability to retain its absolute power.

We are at a pivotal moment in terms of the Catholic women who are storming the Church's gates today. They include the fiercest fighters who have persisted for decades against incredible odds. They also include thousands of new recruits, from Catholic women ministers to fiery advocates for change spurred on by the tragedy of the sex abuse crisis. Together—and separately—these women are fighting for the soul of the Catholic Church, and they will not be moved. Not by a Vatican that threatens them, censures them, or evicts them from their convents. Not by bishops who boycott their speeches or bar them from Church property. This is their story.

As to the form of the story, I knew it would be a narrative, a journey of sorts. But at first I thought I would divide it into two sections—one on power and one on sex. I soon found, however, that sex and power were too deeply intertwined in terms of the Church's treatment of women to parse them out that way. So I chose instead to focus on voice, beginning with a recent Church effort to silence one woman's voice. I then move on to a brief history of how the Church has tried to silence women through time—from the earliest periods of Christianity to the years following Vatican II to the present day—and how Catholic women, undaunted, have persevered.

The sex abuse crisis of 2002 ended one of the Church's most deadly silences, and this book looks at the impact of that crisis on women and reform. That impact includes the emergence of women leaders of new reform groups—one called, aptly enough, Voice of the Faithful—and the struggle between the new-guard reformers and the old-guard reformers to find common ground. The sex abuse crisis empowered female survivors and their advocates to lead the way in turning pain into hope, and they are in this book. Because the clergy sex abuse scandal speaks to a much deeper problem of sexual secrecy in the Church, you will also find the voices of female reformers demanding an end to mandatory clerical celibacy; arguing for women's moral authority on birth control, homosexuality, divorce, and abortion—though that remains a most vexing issue, as I show, even for reformers; and advocating for a new Christian sexual ethic. Finally, the book brings out women's voices in two crucial areas of Church life—in theology, including biblical history, and in the ever expanding array of Catholic women's ministries. As you read, I know the question will arise: Why do these women *stay?* You will find multiple answers to that burning question, too, throughout the book and in the Epilogue.

This story is by no means the whole story. There are many more Catholic women fighting for reform than I could ever fit in this book. And this book is part of a modern history of this movement that has just begun to be written. After all, thanks to feminist theologians, and even the provocative work of novelists like *The Da Vinci Code's* Dan Brown, we are just beginning to hear about the real women of the early Church, like Mary Magdalene. Much remains to be written about the contemporary women who are following in their footsteps.

Part of the reason for this delay is that Catholic women reformers have been ignored not only by the Church but by the secular feminist community as well. To that end, the women's studies in religion program of Harvard Divinity School held an extraordinary conference in 2002 called Religion and the Feminist Movement. It was mounted in response to the general failure within the feminist movement to recognize and respect the women who have been laboring for decades to change the nature of patriarchal religion, and to include their efforts in second-wave feminist history. One after another, extraordinary change-makers—Jewish, Christian, Muslim, Mormon—stepped up to the podium to share their harrowing experiences taking on the most misogynistic and powerful institutions in the world. It reminded me once again that it is sheer folly to think that we can ignore the world's major religions—and the meaning of religion in people's lives—and still change women's place in the world.

And this is certainly not the only way to tell this story. I am a journalist, not an academic or a Church historian. I did not establish a checklist for change-agents and demand that all of the women who would be in these pages—including the more than one hundred I spoke to personally—agree with each item, with one another, or with me. They have diverse viewpoints. I looked for women delivering the most dynamic challenges; for women quietly modeling change; for those with booming prophetic voices; and for ordinary Catholic women stepping up to leadership because they love the Church so much. It is, in that sense, a subjective history.

Finally, I would be remiss if I did not acknowledge that there are many Catholic women—moderate to conservative—working toward what they see as women's best interests in an unchanging Church. They want to hold the line, to keep the pre-Vatican II institution alive. Some even call themselves "papal feminists" because they support Pope John Paul II's vision of gender relations, a vision that maintains the Church's status quo. They have their own books. They are spirited and devoted, too. But they don't represent the future I want to see for the Church.

So I am no innocent bystander. I am a woman with a history, a woman scarred, a woman at her wit's end. I abide the women in this book. I echo their words. I applaud their patience. And I remind this Church how fortunate it is to have such brilliant and devoted women clamoring for the Catholic hierarchy to open its doors, bring the wizard out of the sacristy, rethink the sacred with women in mind, and make a new Catholic Church.

THE REVOLT
OF THE ERIE
BENEDICTINES

When I phoned Sister Joan Chittister before my visit, she was roaring mad. Turns out that the *New York Times* magazine editors had killed a profile about her just before it was to go to press. The time was early 2002, just as the *Boston Globe*'s spotlight team had let loose a fury of pedophilia accusations against Catholic priests. The reason Chittister was given: The story wasn't about pedophilia. This was after the freelance journalist had trekked, the weekend before Christmas, to the shores of Lake Erie in Pennsylvania, just a few miles short of Canada, to spend time with Chittister. And after photographer Joyce Tenneson, producer of a smashing book of photographs called *Wise Women*, went off to take Chittister's picture.

Chittister usually hates pictures of herself. She especially hated the one in *Time* magazine that made her look slightly diabolical, under a headline about her "Dangerous Talk." That was published in August 2001, when Chittister made international news. She had been invited to speak in Dublin, Ireland, at the first international conference of organizations worldwide on the ordination of Catholic women to the priesthood. No sooner had she been invited than the Vatican forbade her to attend. Chittister had a choice: Stay dutifully at home or defy the pope.

Into the *Times'* trash went the story. Lost was a moment of real celebrity that might have awakened people to the extraordinary women leading the progressive Catholic Church reform movement in the United States—women who had just provided a prescient model for dissent. Chittister was incensed. "In the middle of this unprecedented male ecclesiastical scandal," she immodestly intoned, the newspaper of record had failed to see how important it was to bring to the world stage "the women who represent the very best in the Roman Catholic tradition."

Chittister has devoted her life to reaching Catholic women, emboldening them, and moving them to move the Church to change. She entered Mount Saint Benedict Monastery at the age of sixteen in 1952. But her real work began in those misty mornings before the dawn of the second wave of feminism. When many of us were burning our bras, symbolically at least, Chittister and a battalion of nuns at the Erie Benedictine monastery were doing something equally defiant—tossing habits, veils, and starched white collars into one collective garbage heap.

Joan Chittister: The Early Years

When I visit Chittister, she has just turned sixty-six. Of average height and broad-shouldered, she walks with a limp and a lumbering gait, keepsakes of her bout with polio. The disease struck when she first entered the convent and confined her to a wheelchair for nearly two years. Chittister has marble-blue eyes that fill up with tears at a moment's notice. She has short, unstylish salt-and-pepper hair. She dresses practically, in jersey knits in blues and maroons, and sensible shoes. She has a big, robust laugh, and a tough-guy quality about her.

Most of the Erie Benedictines—who numbered 135 in 2002—live in dormitory-style rooms at the sprawling monastery. Chittister is among those who struck out on their own. She lives in a simple, two-story, 150-year-old colonial house in a tired section of downtown Erie. She has three housemates: Sister Mary Lou Kownacki, who runs the Neighborhood Art House; Sister Mary Miller, who operates the Emmaus Soup Kitchen; and the indefatigable Sister Maureen Tobin.

When Chittister entered the order, Tobin was already there. A lumbering presence herself, shorter than Chittister, very plump, and slightly disheveled, Tobin was Chittister's principal when Chittister taught high school English, beginning in 1959. Tobin became her dear friend and, later, her manager. Today, Tobin schedules Chittister's speaking engagements, makes her plane reservations, screens her voluminous mail, picks up her guests at the airport, and keeps everyone who wants a piece of Chittister in an orderly line, getting to them as she is able to, while hardly ever losing her temper.

Chittister's house is full of fabulous old wood—maple, mahogany, tiger oak. On the first floor is a comfortable living room, with two couches, overstuffed chairs, and a color TV on which Sisters Mary and Maureen watch a lot of basketball. A tiny prayer room opens out into a dining room, where a painting of the Last Supper hangs on the wall. The painting is darker and more mysterious than the original. Instead of the all-male gathering imagined by the Church fathers, this one—more realistic, say the feminist theologians—includes children and women. "Who do you think baked the bread and poured the wine?" snapped a visitor, Sister Bernadette, as I stood staring. "You don't think it was those Palestinian men, do you?"

Chittister's study, where she writes (she's authored more than thirty books and counting), is at the top of the stairs. Her chair is too high for her desk—ergonomically speaking—so she has to slouch to peck furiously on her little laptop. Otherwise, she is computer-savvy. She sleeps with a tiny, voice-activated tape recorder under her pillow so she won't miss those elusive moments of nocturnal creativity. She also has a C-Pen—a kind of computerized Magic Marker—that holds up to one thousand words of text that she can then download to her computer. Testament to her work, book jackets hang in frames all around. On her desk is a photo of her mother, her father, and herself as a baby in her father's arms. He died when she was three, an event that left her bereft.

For nuns, two major developments caused cataclysmic changes in religious life. The first was the Sister Formation Movement of the 1950s. A response in part to the new requirements for professional certification, particularly for teachers, American nuns pursued advanced education as never before. They earned master's and doctoral degrees, becoming among the most educated employees of the Roman Catholic Church. The other influence was the Second Vatican Council (1962 to 1965). It called for the "renewal of religious life," empowering religious communities to change everything, from prayer rituals to clothing to internal governance to their place in the world.

When religious renewal reached the Erie Benedictines, back in the 1960s, however, Chittister was appalled. A devout thirty-year-old woman, she had become deeply accustomed to wearing her habit, to the semicloistered life, and to a protected and prayerful existence. She didn't know who she would be with all that gone. She was so upset that she thought about leaving. She talked to her family. She prayed. In time, her personal crisis was resolved. "I always say that my conversion came the day I said to myself: 'Joan, the question is: Am I or am I not a Benedictine in the bathtub?'" She found that she was. She also found that being a Benedictine nun for her represented "a lifestyle . . . a mind-set, a commitment. It was a way of being in the world. It wasn't a way of being *out* of the world."

Deciding to stay meant facing the demands for change squarely. Chittister did. In fact, she became a chronicler of the process, director of research and lead writer of a book called *Climb Along the Cutting Edge: An Analysis of Change in Religious Life*. She also began to speak publicly.

A conversation with a young nun got the ball rolling. Chittister had earned a master's in communication arts from the University of Notre Dame. She was studying for her PhD in communications theory, with a minor in social psychology, at Penn State. The young nun was a student at Penn State, too. They met for a marathon lunch, which lasted until nine o'clock at night.

Chittister talked about what she was learning about social change in her courses, and from the young campus activists whose guiding principle was: Question Authority. By contrast, the basic goal of monastic life had been blind obedience. "The perfect nun," decreed the stern Mother Superior in the 1950s film *The Nun's Story*, "is obedient in all things, unto death." With Vatican II came the liberation of Catholic nuns, an end to a smothering cloud of "sin" under which they all had lived—a category that included useless conversation, too much (or too little) self-flagellation, and even asking a simple question. Chittister explained that what she was learning in her social psychology courses about social change applied directly to the 360-degree life changes American nuns were experiencing. When lunch was over, her new friend said: "Would you come and tell my community what you just told me?"

That question launched Chittister on her life's work. In the early 1970s, she developed a workshop for nuns in the throes of renewal; she called it "Self-Understanding in a Time of Change." She was very clear about what the ideal of "blind obedience" had done to Catholic nuns and what they, in turn, had done to the Catholic children in their care. Enamored of the tools of the social scientist, she toted along to her workshops a "fascism scale" and an "anomie scale" and administered both. "Of course," says Tobin, who participated in those workshops, "we were high on everything."

To the nuns in the workshops, Chittister explained: "Fascism is not a personality characteristic. It's not genetic. It's not chemical. It's not born. It's made. . . . Every time you say to a second grader, 'Put your shoulders back. Put your thumbs on your seams. Don't talk. Everybody right behind the other one,' we're training fascists. We've made a Church full of fascists. You and I have done it. We have to start to reverse it."

Chittister soon began to use the controversial studies of Stanley Milgram in her workshops. As the world was coming off the horrors of Nazism, one question loomed: How could ordinary people have done such monstrous things? At Yale, to help answer that question, young Milgram designed what became a widely replicated social psychology study. His subject: obedience to authority. Milgram's study featured an experimenter directing the activities; volunteer "learners," who had electrodes strapped to their arms; and volunteer "teachers," who operated the consoles that punished bad learners with electric shocks.

"The purpose of the study was to see how many volts a person would be willing to give an innocent bystander when directed by an authority figure," Chittister told her workshop participants. "Most people thought that no one would cooperate up to two hundred volts when they knew someone would be suffering like that. . . . That seemed sensible. But it was wrong." In fact, Chittister explained, the study found from 60 percent to 85 percent of all teachers—who did not know that the learners were plants—obeyed the experimenter and punished the learners to a maximum of four hundred fifty volts—which in real life, would be a lethal electric shock.

Chittister's message about resisting authority was a dramatic challenge to everything the nuns had learned and lived. "That formed me for Dublin years before Dublin," said Chittister, referring to the venue for her confrontation with the Vatican. "I knew there would come a moment in my life when I'd be taken to the console." Would she hurt other people by doing as she was told? Or would she listen to her conscience?

By the age of thirty-five, Chittister had become the youngest prioress ever of the Erie Benedictines. Influenced by the climate at Penn State, she also became an antiwar and antinuclear activist, a passionate commitment she would maintain throughout her life. In her first year as prioress, she and her other nuns led a "die-in" at the cathedral. They dramatized a nuclear explosion, which left the steps and streets littered with the bodies of nuns pretending to be dead. This was deeply disturbing to the vice president of the local phone company, who called to protest and stopped donating funds to the monastery.

Chittister Takes on Feminism

While feminism didn't figure into Chittister's consciousness when the nuns began to toss off the baggage of the past, it did take hold in the years that followed. In the early 1970s, she was invited to speak at a workshop on women's communities in Berkeley, sponsored by the women's journal *Signs*. It was a secular gathering. She wanted very much to go. It was her first time "out."

At the opening of the seminar, a conflict erupted. Appalled at hearing that there was a nun in the room, a young woman seated in a front row insisted to the moderator that Chittister be asked to leave.

Chittister rose. "I can see there's a problem here," she said. The woman did not turn around. "I have to tell you very honestly that I don't understand it. . . . Can you tell me why you want me out of this room when it was so important to me as a woman to come to this meeting?"

"Because you nuns have said, 'Yes, Father' all your lives and taught every generation of women after you to do the same," snapped the woman.

"You're right," said Chittister. "But we're waking up, too. . . . We need your help."

At another point in the session, the woman asked Chittister publicly a question that Chittister by then had asked herself a hundred times. "How can a woman like you stay in the Catholic Church?" "I am a social psychologist," Chittister responded. "I know that symbol systems are a deep part of our identity and our spiritual life as well. I know it is, if not dangerous, certainly difficult after a certain point in life to try to reabsorb a whole new set of symbols and make them your own." You may never feel "at home."

She also told her oyster story. In it, Chittister is the sand and the institutional Church is the oyster. "During the spawning season . . . when the sand invades the oyster, the oyster emits a gel to protect itself from the sand," she explained. "The more sand that comes in, the more gel is excreted. So at the end of the process . . . you have a pearl, [and] the oyster is more valuable." In the oyster analogy, Chittister found her role in the Catholic Church: "I discovered the ministry of irritation."

As the intersection of religious and secular women grew, so did the exposure of nuns to the feminist secular world. So perhaps it was inevitable that eventually, Chittister would discover Judy Chicago.

In 1981, a friend took her to see Judy Chicago's exhibit, *The Dinner Party*, in Cleveland, Ohio. A stunning and massive multimedia presentation, *The Dinner Party* was the 1980s version of the 2000 phenomenon *The Vagina Monologues* by Eve Ensler. Described by promoters as a symbolic history of women in Western civilization, *The Dinner Party* consisted of a giant triangular table covered in linen, forty-eight feet on each side. Displayed on the table were thirty-nine place settings, each honoring a woman in history, real or mythical. Each setting featured an embroidered cloth runner, with stitched images recalling the woman's life; a chalice; and a china-painted plate. It was the plates that caused the stir: Each bore a vivid, three-dimensional, abstract image, and most images resembled female genitalia, rising dramatically out of the plate. The image on Petronitta de Meath's plate, for example (she was burned as a witch in Ireland in 1324), was ablaze in reds and golds and rimmed in flames. Many of the women recalled were religious—from the Biblical Sophia to the Old Testament heroine Judith; from Saint Bridget to the nun and mystic Hildegarde of Bingen to the German poet and playwright Sister Hrosvitha. The glistening, ceramic-tiled "Heritage Floor" was inscribed with the names of 999 more women, restored to history.

Chittister loved it. She loved it so much that when she got back to the monastery, she rented three buses to take all of the Erie Benedictines to Cleveland to see it. That wasn't enough. Next she worked with a committee to encourage each chapter of the Federation of Saint Scholastica to create their own *Dinner Party*, designing runners and place settings to symbolize their communities and the women in them.

That done, on a summer day in 1982, some one hundred delegates from the federation's twenty-three member communities met at the Mount Saint Scholastica Monastery in Atchison, Kansas. In a specially designated room, they set up their own triangular structure to display their runners and place settings. Symbols abounded: circles, stained glass windows, hands, windmills, and for the Erie Benedictines, a burning bush. Together, they watched a film about *The Dinner Party*. Chittister used the opportunity to lead a discussion about the vagina as "a potent building symbol" that is seldom recognized that way. Then the nuns moved past their vaginas to talk about the general failure to be recognized for their contributions, particularly in their own Church. The excitement built, as did the hope for change. In a great spirit of solidarity, the nuns wrote the names of beloved federation members across the floor.

Then, Chittister tells me, with obvious glee, they pushed the desks back, rose to their feet, held hands, and sang: "I am Woman!" "We loved it," says Tobin, smiling broadly, face flushed. "It radicalized me. It radicalized the community." Two decades later, the Erie Benedictines decided to be radical again.

Church Target

In the years that followed, Chittister began to write prolifically and speak before larger and more diverse audiences all over the world. By the dawn of the twenty-first century, she had spoken in thirty-nine countries and become the "free-floating star" of the Catholic Church progressive reform movement. Her speaking engagements had to be booked three years in advance. The book *Spiritual Questions for the Twenty-First Century*—with a host of illustrious contributors, from peace activist Father Daniel Berrigan to Bishop Thomas Gumbleton, from Notre Dame professor Richard P. McBrien to theologians Elizabeth Johnson, Diana Hayes, and Ivone Gebara—was published in her honor.

Chittister developed a radical perspective on Catholicism and change. While she came to the wider public's attention over the issue of women's ordination, her real mission is much broader than that: She is out to end the "clerical culture" of "exclusion" and "domination" to transform the Roman Catholic patriarchy. At the heart of her reformist agenda is the complete acceptance of women as equals in the Church. She demands gender-neutral prayer language and a redefinition of the very nature of God, away from the exclusively male. Within the Church hierarchy, Chittister had become a force to be reckoned with.

Late in 2000, she brought her ire to the Vatican's backyard, in a speech about feminism and patriarchy. "We are . . . descendants of . . . a Roman Church who built a bad theology of male superiority on a bad biology . . . who taught that women were inferior by nature and deficient of soul—the servants of men and the seducers of civilization," she boomed to her audience of one thousand members of religious orders, seated and standing in the aisles, in one of Rome's largest auditoriums. "The notion that God would create women with brains in order to forbid them to use them paints God as some kind of sadist. The notion that the God of Mary and Eve . . . trusts more to men than to women in the divine plan for salvation ignores the very place of women in the Christian tradition."

While that speech was being translated simultaneously into five languages, back in the United States, trouble brewed. The day after her speech, a call came for Chittister in the convent where she was staying. The caller was a Pittsburgh stringer for the independent newspaper the *National Catholic Reporter*, on whose board she then sat. She took the call in a pantry. "Don't ask me why, but in every mother house in Europe, every telephone is in the pantry," she recalled. "So I'm standing in this pantry with the cleaning supplies . . . with this lousy phone [listening] to this reporter." The stringer said that a memo had gone out from the diocese of Pittsburgh (which is near Erie) to hundreds of Catholic school teachers, refusing to pay for them to attend the upcoming annual conference of the National Catholic Education Association (NCEA). The reason: Sister Joan Chittister would be a keynote speaker.

In March 2001, Chittister returned from her writing retreat in Europe to her home in Erie. By then, five dioceses around the country had joined the boycott of NCEA's April conference—a generally mainstream, noncontroversial activity. But while the diocesan administrators supported a boycott, their employees did not. The national order of the Sisters of Mercy set up a scholarship fund to cover the costs for any teacher who wanted to attend whose diocese refused to pay. The Erie Benedictines launched a postcard campaign in Chittister's defense. And both the Thomas Merton Center and the Sisters of Divine Providence in Pittsburgh publicly announced their choice of Chittister to receive their annual peace and justice awards.

As Chittister was gathering her courage to go to the NCEA conference, she got hit with another missile from the Church hierarchy's arsenal—this time from the Vatican. She had agreed to speak in Dublin in June, at the first international conference sponsored by Women's Ordination Worldwide, with the help of an Ireland-based group, Brothers and Sisters in Christ (BASIC). To her astonishment, the Vatican sent a letter to her monastery forbidding Chittister to speak. If she did, it said, there would be "grave penalties." By that time, the Vatican's ban on women's ordination to the priesthood had been declared definitive teaching; even discussion of the subject was forbidden.

Painful as that possibility was, Chittister knew, no matter what, she would go to Dublin. The issue was not ordination but silencing. The Church teaches that the voice of the laity—the *sensus fidelium*, or sense of the faithful—is essential to the development of Church doctrine, to ensure, as Chittister describes it, that the doctrine "makes sense in the lives of the faithful." As a nonordained Catholic, Chittister and all nuns consider themselves part of the laity. So when she was told "to make a decision that was for the good of the Church," she says, she decided she must go to Dublin because, to her mind, silencing has never been good for the Church. On an even more personal level, Chittister felt: "My integrity depended on it."

But ominous cautions came from Church reform friends. "They told me: 'It's too small of an event. It's not worth it. . . . If you do it, they'll trash all your work. You'll never speak again. . . . We really need you. Don't do it.'" Chittister held firm. "If I *don't* go, my work is over," she told them. "I'll never write another word. I'll never speak another phrase. . . . How could I ever look at another woman . . . and say, 'You're in an abusive situation,' when she looks back and says, 'Joan, you were abused, and you took it.' . . . Maybe one significant act of conscience, however small, is worth more than all the books and all the words ever written, ever said."

Chittister boarded a plane for the NCEA annual meeting in Milwaukee. The tension was so high over the boycott that the NCEA assigned her bodyguards. Flanked by those guards, Chittister climbed the stairs to the convention center to give the closing keynote address on April 20. She received a rousing standing ovation before she even said a word.

"When your most sublime ideas meet the greatest resistance, remember that today's heresy is tomorrow's social dogma," she told some fourteen thousand Catholic school teachers and administrators who attended from the nation's dioceses—several thousand more than had been anticipated. "So it was when Galileo questioned the nature of the universe. So it was when Luther asked for the publication of the scriptures in the vernacular. So it was when Sojourner Truth demanded an end to slavery, so it was when Elizabeth Cady Stanton went on a hunger strike for a woman's right to vote, so it was when John Courtney Murray argued for freedom of conscience. . . . Teach them to question . . . for all our sakes, teach them to think." It was the same message she had brought to the small convent of nuns at her first workshop decades before.

Chittister left NCEA, got on a plane to Iowa, and spoke there the next day. Two days later, her alma mater, Mercyhurst College, had a reception in her honor. The day after that, she "completely collapsed." She was rushed to the hospital, diagnosed with anemia, pneumonia, and a raging infection. She had to have gynecologic surgery. She lay in a deeply weakened state for one month.

Chittister had barely recovered from that ordeal when it was June, time for her to go to Dublin for the Women's Ordination Conference. Still exhausted, she prepared for the trip, the Vatican's threats hanging over her. Up until the night before her flight, Chittister was sure she would be acting alone in defiance of Rome. She was also sure that the consequences for her would be dire, that "everybody would fade into the darkness and I'd be left there . . . just one more scapegoat." She thought, "My life will be over."

The Path of the Prioress

Like Chittister, Christine Vladimiroff, the prioress of the Erie Benedictines, stood in the Vatican's line of fire. It was Vladimiroff, in fact, who received the letter from the Vatican's Congregation for the Institutes of the Consecrated Life. It ordered her to issue a "precept of obedience" to Chittister, forbidding her to go to Dublin. If Vladimiroff did not do that, said the letter, there would be "grave" penalties. Those penalties could range from excommunication of both women to dispensing them from their vows, banishing them from their home, or even disbanding the Erie Benedictine community altogether. Chittister pleaded with Vladimiroff to protect herself and the other nuns and issue the precept.

Vladimiroff, too, has lived her whole adult life as an Erie Benedictine. Unlike Chittister and Tobin, she has a formal appearance. Her hair is combed perfectly into a 1950s pageboy. She wears neat tailored suits. White blouses. Pumps. And there is also her manner. Vladimiroff has a sternness about her. When I meet her, she shakes my hand but gives me a hard and scrutinizing look. She reminds me of my eighth-grade teacher, Sister Elaine. Both of them scared the living daylights out of me.

As prioress of the Erie Benedictines, Vladimiroff lives at Mount Saint Benedict Monastery. She has a modest, no-frills room as well as a long, bright office. She also has a friendly, pitch-black cocker spaniel, who she kept despite a rule against having pets. When I asked how she got around that, she simply replied: "I never asked for permission."

Reflecting the dramatic decline in the flow of young women entering convents in the United States at the end of the twentieth century, most of the 135 Erie Benedictines in 2002 are elderly. At any one time, at least ten of them near to death are cared for in the building's infirmary. While there have never been more than two hundred Erie Benedictines, now just two or three new postulates join each year.

By day, the monastery's stone halls are dim, but by night, they are dark and dreary. Little night-lights are spaced equal distances apart, to save on electricity. The old carpeting squishes under your feet. The halls are lined with paintings of Saint Benedict and his sister, Saint Scholastica, namesake of the Erie Benedictines' federation. There are no bloody crosses with a hanging Jesus, but rather the welcoming and peaceful risen Christ, in flowing robes.

Under Vladimiroff, the Benedictines have become a greater force for community improvement than ever before. As she explains it, they "direct the Neighborhood Art House, teaching music, dance, pottery, and painting to the poorest children in Erie . . . run the largest soup kitchen and pantry in Pennsylvania . . . administer the welfare-to-work program in six counties . . . operate a day-care center for infants and children of migrant workers . . . direct a community center where children with physical and mental disabilities play, and run seventy-five units of housing for the poor elderly." They also teach environmental education, are participants in the peace movement, and offer a host of retreat programs "for the nurturing of spirituality." Vladimiroff herself has been a major advocate for world hunger relief, once heading the Second Harvest Food Bank Network and serving on the presidential Food Security Advisory Committee.

When the letter from the Vatican arrived in March—three months before the Women's Ordination Worldwide conference was to be held— Vladimiroff was deeply saddened. She saw the Vatican's request as a blatant affront to the Erie Benedictines who were "trying . . . to have a voice in the

Church." Like Chittister, it wasn't that the conference concerned the Vatican's opposition to women's ordination that pained her so. It was that she could not "order something I was in total disagreement with, and that is silencing."

As followers of the 1,500-year-old rule of Saint Benedict, which dictates every aspect of monastic life, the Erie Benedictines have very definite guidelines about obedience and authority. "Obedience in Benedictine life is not a matter of someone in authority telling someone to do a task or not to do a task," Vladimiroff explains. "Benedictine authority and obedience are achieved through dialogue between a [community] member and her prioress in a spirit of coresponsibility. . . . The role of the prioress in a Benedictine community is to be . . . a guide in the seeking of God. . . . It is the individual member who does the seeking."

Vladimiroff spoke with Chittister many times during those days of deliberation. But not just with Chittister. She met with the entire community. With sisters in small groups. With nuns individually. She told them the decision was hers, but she had no plans of making it until she had heard from every member. She sought the advice of bishops, religious leaders, canonists, and other prioresses.

She also walked smack into the lion's den. Rather than roll over in the face of Vatican intimidation, Vladimiroff attempted the unthinkable in recent Catholic Church history: She initiated a dialogue with Rome. She wanted the committee that had sent her the letter to clarify the basis on which they were asking for her to silence a member. She also wanted to prevail on them to let the Benedictines handle the matter the Benedictine way, through "a process of discernment" that would have respected Chittister's spiritual journey rather than slamming down an absolute and, Vladimiroff felt, unconscionable edict.

Vladimiroff asked for a meeting and got it. Just a few weeks before the Dublin conference, she boarded a plane and headed for the Vatican. She took with her two canon lawyers—one an Erie sister and another a priest. For two and a half hours on May 28, they met with representatives from the Congregation for Institutes of the Consecrated Life, a committee of four made up mostly of men. "I was able to lay out my reasoning and my feeling for what needed to happen, and they laid out their reasoning. . . . No one hid the fact that there was a disagreement, but I didn't feel put down," she reported. Before the Erie contingent left, the committee asked Vladimiroff to put in writing what she thought.

Vladimiroff came home, wrote another letter arguing against the Vatican's ultimatum, and sent it. But as the day for Chittister's trip drew closer, no word came from the Vatican. Its threat stood. And the pressure on Vladimiroff to make a decision mounted. Some of the sisters felt obligated to abide by the directive "because the Holy Father's asking us to." Others were so upset that they wanted to "get out of the Church" entirely. And there were nuns who felt everything in between. Vladimiroff herself kept vacillating, rationalizing both sides in her head. In truth, she was "scared to death" at the thought of actually refusing to issue the silencing order.

Vladimiroff went "deeper into prayer. . . . I went through Lent and Holy Week like I've never done before in my life. . . . I became totally attentive to all the things I said this life was about and my faith was about, and [how] I have to live it." To her, the Church had veered far away from the message of the Gospel Jesus, of equality between women and men, rich and poor, cleric and layperson, nations and religions, of the Jesus who was "the way, the truth, and the light." The truth, she said, "must withstand dialogue."

But the "terror of the whole thing" for Vladimiroff concerned what would become of her community. If she refused to issue the precept and Rome removed her as prioress, she would no longer be able to protect them. "I'm sixty-two," she said. "I've been here since I was seventeen years old. . . . It's my whole life. . . . These women raised me. . . . To see people in their eighties moving toward death wondering will the community be there . . . that is what saddened me."

It was June 21, the night before Chittister was to board a plane for Ireland. An hour before evening vespers, Vladimiroff finally had the text of the letter that she had written a hundred times ready to fax to Rome. She turned it over to the five members of her advisory council and told them she wanted to read it to the entire community.

The Erie nuns filed into the A-framed chapel, with its soaring beamed ceiling and looming stained glass windows in brilliant reds, purples, blues, and greens. They settled into wooden pews before a simple altar covered with a white linen cloth. Nobody knew then what Vladimiroff's decision was—not even Joan Chittister.

"After much deliberation and prayer, I concluded that I would decline the request of the Vatican. It is out of the Benedictine, or monastic, tradition of obedience that I formed my decision. There is a fundamental difference in the understanding of obedience in the monastic tradition and that which is being used by the Vatican to exert power and promote a false sense of unity inspired by fear," Vladimiroff said in a public statement.

"Sister Joan Chittister, who has lived the monastic life with faith and fidelity for fifty years, must make her own decision based on her sense of Church, her monastic profession, and her own personal integrity. I cannot be used by the Vatican to deliver an order of silencing. I do not see her participation in this conference as a 'source of scandal to the faithful' as the Vatican alleges. I think the faithful can be scandalized when honest attempts to discuss questions of import to the Church are forbidden."

Vladimiroff walked to the altar, turned to the women, and read the text of her letter. When she had finished, she placed the letter on the altar. "Anybody who feels in conscience that they can come forward and put their name on this paper, I would really like you to do that," Vladimiroff said. Then she walked to the front pew and sat down.

Vladimiroff had no idea that rows of women were lining up behind her. On shaky legs, with walkers, and in wheelchairs, they made their way down all four aisles. Though some, like Sister Maureen Tobin and Sister Bernadette, were scared, they signed the letter anyway, doing so on the same altar where they had taken their vows as nuns half a century before. The council members took the papers up to the infirmary—everything from the chapel is piped into the infirmary—and invited those members to sign on. And to the nuns who were out of town, the council members faxed a copy of the letter to see if they wanted to participate. In the end, 127 of the 128 active members of the Erie Benedictines cosigned Vladimiroff's letter.

Meanwhile, all over the country, women in charge of the twenty-two other monasteries of Benedictine nuns in the Federation of Saint Scholastica—some who had made their own *Dinner Party* settings and belted out "I

Am Woman" decades before—were putting the finishing touches on their own letter of support.

When it was all over, "I went up to the microphone—Christine called me up," recalls Chittister. "The community has a custom of blessing sisters who are going into dangerous places. When they go into the Third World, where we're worried about our sisters, the whole community stands, they put their hands up, and they sing. . . . [That night] they blessed me, and then I put my hand on Christine's head and the community sang a blessing. I said to the community: 'I want you to know that you have just witnessed one of the most prophetic acts I have ever seen. . . . I don't want you to miss it because you're it. You're it.' "

A Model for Change

In the Roman Catholic Church, to say no to authority that claims to derive from God is a momentous act. It is a Church where not only actions but thoughts are policed, where sin begins in the mind. It wasn't just the nuns in the convents who lived in a world where opportunities for sin abounded; where the last word came from a hierarchy claiming to be far holier than they were; where individual voices and consciences were smothered and silenced. It was all Catholics. Which is why the courage of the Erie Benedictines meant so much to so many.

The next day, as the faxed letters arrived in Rome, Chittister boarded a plane for Dublin. Peripheral neuropathy had numbed her feet, and when she landed, she walked warily on a cane. She made her way to the convention center and took her place on the dais, before an audience of nearly four hundred participants from twenty-seven countries. "I walked into that conference room, at that college, into the midst of the most sensible, intelligent, sound, sane, and simple women I've ever seen," she said. "This was not the lunatic fringe of Roman Catholicism. . . . This was the solid, sincere, and holy center of the Church . . . a collection of serious and seriously committed women who believed, who really knew, they had a call to ministry. It was a marvelous moment."

Chittister spoke about the desperate need for women disciples to save the Church in what she called a "priestless period." She issued a clarion cry for revolt, calling on women to become disciples of Christ, right here, right now, no matter the price. "Christian discipleship . . . is dangerous," she warned. "To follow Jesus . . . is to follow one who turns the world upside down, even the religious world. . . . True discipleship says the truth in hard times. . . . To the true disciple, the problem is clear: The church must not only preach the Gospel . . . it must demonstrate what it teaches. . . . The church that preaches the equality of women but does nothing to demonstrate it within its own structures . . . is out of sync with its best self and dangerously close to repeating the theological errors that underlay centuries of Church-sanctioned slavery." She never called specifically for women's ordination to the priesthood; her demand, as promised, was for dialogue.

And Chittister held the hierarchy's feet to the fire. "Men who do not take the woman's issue seriously may be priests, but they cannot possibly be disciples," she said. "The people need more prophets of equality, not more pretenders to a priesthood of male privilege." Noting the successful struggles in Hinduism, Buddhism, and Judaism for women's spiritual equality, she said: "Only in the most backward, most legalistic, most primitive of cultures are women made invisible, made useless, made less than fully human, less than fully spiritual. The humanization of the human race is upon us. The only question for the church is whether the humanization of the human race will lead as well to the Christianization of the Christian church." Raging and rebellious but full of love for Christ and his message of equality, she called for change and courage and a renewed Catholic Church.

The Vatican never made good on its threats. Most people think they finally realized who they had tried to take on. "All those women, gutsy, bright, brilliant women, speaking the truth, that was a watershed event," said Linda Pieczynski, spokeswoman for Call to Action, among the nation's oldest and largest progressive Church reform organizations. "They basically dared the Vatican to do to them what they were going to do to Joan. By sticking together in that great number, the Vatican blinked. It just shows you what can happen with solidarity."

American Catholics long ago stopped agreeing with the Church hierarchy on myriad matters, from mandatory celibacy to women's ordination, from contraception to homosexuality to divorce. But their rebellion has been silent. Some left the Church. Others stayed and pretended. The Erie Benedictines showed what could happen if Catholics gave voice to their revolt—together. They provided a model for a new spirit of Church reform, inspiring many of the Catholics who clamored for change in the wake of the U.S. priest pedophile crisis of 2002.

Not just the Catholic Church but any institution that behaves autocratically counts on silence to keep its power. Change only comes when people take back the right to speak the truth. Social change begins with voice. The Catholic clerical hierarchy is right to fear that challenge most of all.

2

✝

THE CHURCH AND REFORM: A WOMAN'S PLACE

The history of women's place in the Catholic Church is one of fits and starts, of rising power followed by backlash, of emerging authority squelched and denounced by a threatened all-male hierarchy. It happened in the ancient Church; it happened in the Middle Ages; it happened in the centuries that followed; and it's happening now. What is amazing is that Catholic women keep coming back, pushing for voice and empowerment, demanding full participation, advocating and making change in a Church they continue, against the odds, to love.

The roots of the women reformers' commitment to the Church run deep. They embrace Jesus Christ as undeniably a friend of women, a man who they believe "violated almost every known cultural norm that kept women oppressed, suppressed, and invisible," according to activist Sister Christine Schenk. Indeed, as the Bible records, women journeyed with Jesus "through cities and villages" at a time when women were barely allowed to talk to men in public. When a bleeding woman touched the hem of Jesus' garment, despite the taboo that declared her unclean and unfit for human contact, he looked on her with compassion and healed her. Jesus defended the right of a woman to listen with the men to his preaching, instead of cleaning house. When presented with a woman publicly accused of adultery to judge, he refused to condemn her and forgave her instead.

In return, women devoted their lives to Jesus and his ministry. They provided financial support. They remained at the foot of his cross after frightened male apostles fled. Women anointed and buried Jesus, and it was women who saw the empty tomb and the risen Lord first, making them the primary witnesses to the heart of the Catholic faith, the Resurrection. After Jesus' death, women ran house churches. They were teachers and prophets, disciples and deacons, apostles, evangelists, missionaries and martyrs. In the last chapter of Paul's letter to the Romans, ten of the twenty-nine church leaders he commends are women.

But the early Christian Church was under great scrutiny. The public involvement of women drew derision and unwanted attention. One of the ways the early Christians chose to defend against that derision was to put Christian women back in their place. The fate of Mary of Magdala—aka Mary Magdalene—is a case in point. The most famous prostitute in Church history was not really a prostitute at all but a devoted disciple and the first witness to Jesus' resurrection, which led the Church fathers to call her the "Apostle to the Apostles."

Mary of Magdala's slide from Jesus' cherished companion and benefactor to the Church's mythic "fallen woman" was facilitated by Pope Gregory the Great in the sixth century. He interpreted Magdalene's seven demons, cast out by Jesus, as "vices" instead of illnesses, and collapsed the several stories of women named Mary in the Bible—including a repentant sinner— into one. Clearly propelled by the Church's misogyny, Gregory chastised Magdalene for her "forbidden acts" and scandalous displays, for the "proud things" she said, and for her "crimes." She was redeemed from this criminal state, maintained Gregory, by atoning "for every delight . . . in order to serve God entirely in penance." It was feminist Biblical scholarship that ultimately restored Magdalene's reputation and her place as a leader in Christian history.

Eventually, women deacons were renamed deaconesses and demoted on the power ladder, until they were completely excluded from that ordained ministry. One twelfth-century authority on canon law attributed this early demotion to women's "monthly defilement," which required her removal from the diaconate and from the altar. Second only to priests in spiritual power, ordained deacons can baptize, marry, and bury Catholics—an office from which Catholic women remain excluded even today. In time, women lost what power they had to speak in church, to preach, and to teach or hold any authority over men. Some women literally were erased, like the apostle Junia—who appeared in the New Testament—whose name morphed into Junius, a man.

These and many other demotions in power and authority were supported by a misogynistic theology that demonized women and female sexuality. Saint Thomas Aquinas described a female as "a misbegotten male," with a defect in her reasoning ability; he saw "man as the beginning and end of woman, as God is the beginning and end of every creature." Augustine believed that man alone was made in "the image of God," and saw woman's subordination as a manifestation of the order of nature and divine law—not so different, pointed out one feminist theologian, from the Church's rationalization for slavery, which was based on the belief that slaves were considered "by nature not free" and so by nature "intended for servitude." The early Church father Tertullian saw Eve in all of us, describing women as "the devil's gateway."

In the Middle Ages, women rose again in power and influence. Billed as the "first women's movement," the Beguines, for example, were free-spirited lay Christian women who lived from the twelfth to the fourteenth centuries in an independent spiritual community. They spent their days ministering to the sick, the poor, and the spiritually hungry. As their popularity grew, the Church became increasingly threatened, eventually condemning them as heretics, taking away their property, and burning at least one of them at the stake.

Continuing the practice of the early Church, women throughout the Middle Ages were wives to priests, bishops, and popes, and mothers to their children. In that sense, at least, women remained intimates in the life of the

Church. All that ended in the twelfth century, when the Church mandated clerical celibacy—defined as the renunciation of marriage but broadly understood as no sexual activity—with a vengeance.

The demand for celibacy grew out of clerical concern over the loss of valuable Church property to relatives of clerics. But as feminist theologian Uta Ranke-Heinemann so vividly documented, it also reflected the Church patriarchs' deepening misogyny. Ambrose, another early Church father, said that priests who continued having children while acting as priests "pray for others with unclean minds as well as unclean bodies." In the seventh century, Pope Gregory I ordered married priests to "love their wives as if they were sisters and beware of them as if they were enemies." In the eleventh century, Pope Urban II gave permission to turn clerics' wives into servants; an archbishop authorized their imprisonment; a London synod made them the property of the bishop; and a prominent cardinal said: "Young husbands, just now exhausted from carnal lust, serve the altar. And immediately afterward they again embrace their wives with hands that have been hallowed by the immaculate Body of Christ. That is not the mark of a true faith, but an invention of Satan."

In 1139, at the Second Lateran Council, priests' marriages were officially declared invalid and celibacy imposed. Yet many clerics—including eight subsequent popes who fathered children—refused for centuries to abide. The consequences for women were dire. Popes referred to priests' wives "without distinction" as "concubines," "adulteresses," and "whores." In some locales, they were refused a church burial. As late as the seventeenth century, a Church synod authorized that it would "inspect the houses of those under suspicion night and day and have the shameful persons publicly branded by the hangman." Concurrently, in Bamberg, the bishop appealed to secular authorities to go to the rectories and "fetch out the concubines, publicly whip them, and place them under arrest." Besides causing women so much pain, the imposition of celibacy violated a basic tenet of the tradition at its healthiest, which is that celibacy should be a freely chosen state.

Up until the eighteenth century, abbesses—nuns heading up convents or abbeys—exercised a great deal of clerical power. They gave out licenses to priests to celebrate Mass in their churches. They were authorized to suspend the priests who were their subjects. They ended excommunications; set up new parishes; proclaimed the Gospel; listened to confessions; and preached publicly. But eventually they, too, were stripped of their powers.

Of course, there has been progress. Many women are giants in Church history. It took centuries but Catherine of Siena, the fourteenth-century Church critic, theologian, and social reformer, and Teresa of Avila, the sixteenth-century spiritual writer, contemplative, and Carmelite reformer, were both named doctors of the Church in 1970. The fifteenth-century French girl-warrior Joan of Arc, after being burned at the stake as a heretic, was five hundred years later canonized a saint. Mary Ward died in 1645, abandoned by her Church as a "heretic, schismatic, and rebel to the Holy Church," for daring to found a religious order of noncloistered women, free to work in the world. Two centuries later, her religious order (the Institute of the Blessed Virgin Mary) and Ward herself were welcomed with honor into the Catholic fold.

In addition to the Sister Formation Movement, which revolutionized the lives of Catholic nuns like Joan Chittister and the Erie Benedictines, the mid-1900s saw the emergence of a Catholic lay movement that began to focus attention on the responsibility of the laity for the Church, in which women played an active organizing role. The fight of Catholic women to change the Church in a way that would improve their own status, however, came later, inspired by two seminal events: the Second Vatican Council and the American feminist revolution of the 1970s. In retrospect, that is when the Church began to lose its hold on Catholic women. It is when the dam broke, when the flood of energy for women's liberation invaded the hallowed halls of the Vatican, inspiring a siege mentality from which the Vatican has yet to emerge.

The Modern-Day Reform Movement

Actually, the Second Vatican Council (1962–65) was no bastion of gender equality. Despite the spirit of *aggiornamento*—literally, "to bring up to date"—that was breathed into it by Pope John XXIII, no female face appeared in the sea of several thousand men at the first session. No women came to the second. Invitations were finally extended to the third and fourth sessions, to a total of twenty-three women, as auditors. No one was allowed to speak.

The council did boldly recognize the sea change under way in women's roles with the second wave of feminism. In one of its enduring documents, the *Pastoral Constitution on the Church in the Modern World*, the council insisted that "with respect to the fundamental rights of the person, every type of discrimination, whether social or cultural, whether based on sex, race, color, social condition, language, or religion, is to be overcome and eradicated as contrary to God's intent." It went on to admit that for women, "fundamental personal rights are still not being universally honored." Among the rights denied, it listed the "right to choose a husband freely, to embrace a state of life or to acquire an education or cultural benefits equal to those recognized for men."

Another document, *Pacem in Terris* or *Peace on Earth*, written by Pope John XXIII during that time, held Catholics responsible for not only recognizing their rights but for claiming them, a stance that would provide change-makers with a strong moral argument for reform. It read in part: "The possession of rights involves the duty of implementing those rights, for they are the expression of a man's personal dignity. And the possession of rights also involves their recognition and respect by other people." While applicable to women in general, the document also spoke to women specifically. "Since women are becoming ever more conscious of their human dignity, they will not tolerate being treated as mere material instruments, but demand rights befitting a human person both in domestic and in public life."

The council ended the Church's medieval insistence that Catholicism be the religion of the state, opting instead for a long-overdue commitment to religious tolerance and freedom of conscience. That, in turn, laid the groundwork for a new independence in Catholic laity that would influence personal morality and sexual behavior in the decades to come. There were critical conceptual changes, too. Paramount among them was a dramatic shift from identifying the Church solely with the clerical hierarchy to an unequivocal recognition of the Church as everyone, clerics and laity, described as "the people of God." And Vatican II greatly expanded the role of the laity—the *sensus fidelium* or sense of the faithful—by recognizing that the faithful had to receive teachings for them to be valid. Indeed, Vatican II held that the very concept of the Church's "infallible" teaching authority—i.e., its inability to err in matters of belief, which was granted to the pope at the First Vatican Council in 1870—had to include the lay faithful. This unerring quality must be manifested by the whole Church, read another crucial Vatican II document.

In the wake of the council meeting, a new fervor for change in the Church swept the land. Women helped to create a network of progressive reform organizations devoted to realizing what they saw as the vision of Vatican II. Among the goals of the developing progressive reform movement were women at all levels of ministry and decision-making; married clergy; optional celibacy; acceptance of homosexuals, the divorced, and the remarried; lay involvement in Church governance, including the selection of bishops; lay participation in the development of teachings on human sexuality; new forms of culturally relevant liturgy, language, and leadership; nonsexist worship language; academic freedom at Catholic universities; ecumenical dialogue aimed at the reunion of Christian churches; and financial accountability. All this was in addition to promoting the Church's historical commitment to peace and social justice.

Janet Kalven, a Jewish convert who attended the Second Vatican Council as a typist, became a leader of the Grailville community of Catholic laywomen in Ohio. Under her spirited guidance, Grailville hosted a groundbreaking feminist conference in 1972 on "Women Doing Theology." It inspired Catholic women to claim spiritual authority and define theology, drawing on their own life experiences. Among the participants in Grailville's early years were the members of the veritable trinity of Catholic women who all but invented feminist theology: Rosemary Radford Ruether, arguably the best-known and most widely read feminist theologian in the world; Elizabeth Schüssler Fiorenza, who would set a new standard for biblical research; and philosopher and trailblazer Mary Daly, who had a knack for shaking things up. After the publication of Daly's classic, *The Church and the Second Sex* (1968)—which took an unflinching look at the Church's pervasive misogyny and its impact on women—Boston College tried to fire her from her position as a theology professor. That inspired months of protests by her male students, which convinced the school to back down. When Daly was invited to be the first woman to preach at Harvard Memorial Church, she decried patriarchal religion from the pulpit and led a mass exodus out of the service.

In 1974, eleven female Episcopal deacons—the so-called "Philadelphia 11"—forced open the doors to the Anglican priesthood. Four courageous retired or resigned Episcopal bishops conducted the "irregular" ordinations before an audience of 1,500 in North Philadelphia's racially diverse Church of the Advocate, symbolically linking a church with a historic commitment to civil rights with the fight for women's rights in mainstream religion.

The action inspired the formation of a women's ordination movement in the United States. Word of the first conference spread like an oil spill, and what had been anticipated as a modest turnout became a flood of 1,200 who made their way to Detroit in 1975. Nuns and laywomen attended (technically, nuns are laywomen but for practical purposes, I'm considering them as distinct), though the nuns who spearheaded the Catholic feminist-movement outnumbered laywomen by a ratio of almost five to one.

Among the speakers was Elisabeth Schüssler Fiorenza, a married German émigré, then associate professor of theology at Notre Dame University, who had attended Grailville's Women Doing Theology conference. Drawing on the biblical notion of "the priesthood of all believers," made up of all the baptized, she argued for a broad concept of ordained ministry, which would come to involve many different functions, all equally valued. "Those women, who as teachers, theologians, assistant pastors, religious educators, counselors, or administrators already actively exercise leadership in the Church, have to insist that their ministry is publicly acknowledged as 'ordained ministry,'" she said. Indeed, for her and many in the early American movement for women's ordination, the goal was never simply to ordain women to the existing priesthood, that is, to "add women and stir," as theologian Mary

Hunt put it. They envisioned a "renewed priesthood" that would not be top-down, would not require celibacy, and would constitute what Schüssler Fiorenza came to call "a discipleship of equals."

That first conference sent Catholic women's hopes soaring. The following year, 1976, the Women's Ordination Conference was officially born. That same year, the American bishops inspired an eruption of interest in progressive reform and provided a model for a participatory Catholic Church. After an unprecedented two years of hearings and parish discussions nationwide, involving some 800,000 Catholics in more than one hundred dioceses, the National Conference of Catholic Bishops hosted the first national assembly of the American Catholic community to set the Church's agenda for justice. More than 2,500 people—official delegates and observers, including lay Catholics, nuns, priests, and bishops—arrived in Detroit for the Call to Action conference, chaired by Cardinal John Dearden.

Recommendations abounded, voted on by lay and clerical delegates alike. They included the easily agreed upon subjects, like nuclear disarmament, workers' rights, and hunger relief. But they also included recommendations on the most controversial issues facing the Church, from women's ordination to the ordination of married men, from freedom of conscience on birth control to freedom from discrimination for gays, from local participation in electing pastors and bishops to open theological discussion, to holding the Church financially accountable. Though the bishops would eventually abandon what Dearden had described as "the new way of doing the work of the Church in America," that conference sowed the seeds for a lively independent reform organization: Call to Action. Spearheaded in Chicago by Dan and Sheila Daley, Call to Action would set itself apart by embracing many issues, adopting a broad Vatican II agenda for change.

In those early days, Catholic women could taste victory. In 1978, there was a second conference on women's ordination, even bigger than the first, attended by nearly two thousand. Besides the United States, panelists hailed from Mexico, Paraguay, Uganda, and India. That conference provided a public venue for a bold, some would say blasphemous, action. On Sunday morning, while a priest celebrated Mass for most of the participants, some 250 attendees retreated to a private room. There they had what conference coordinator Dolly Pomerleau called a "breakaway liturgy," and what the *Washington Post* called "a rebel worship service"—that is, they celebrated the Eucharist themselves.

In 1978, the world got a new pope—the charismatic Karol Wojtyla, Pope John Paul II—and a young Sister of Mercy named Theresa Kane got a new job. A Bronx native, the forty-one-year-old Kane had moved swiftly up the administrative ranks. She had been elected president of her national order and the Leadership Conference of Women Religious, the nation's largest association of leaders of congregations of American nuns, who numbered some 100,000 at the time. An activist for women's ordination, Kane had led the Mercy Sisters and the Leadership Conference in advocating for wider roles for women in the Church.

In 1979, Pope John Paul II made his first trip to the United States. On October 7, he was to hold a Sunday morning prayer service at the Shrine of the Immaculate Conception in Washington, D.C. The nuns in the area were invited to attend. They, in turn, needed someone to issue a "greeting." The charge fell to Kane.

On the day of his visit, some five thousand nuns filed into the shrine. Standing at a podium, Kane expressed solidarity with the new pope's deep concern for the world's poor. She publicly acknowledged the long history of U.S. women religious in building the Catholic Church's system of social, educational, and health services. She asked the pontiff to be "mindful of the intense suffering and pain, which is part of the life of many women in these United States." And she issued what was at once a rebuke and a plea for change: "*As women we have heard the powerful messages of our Church addressing the dignity and reverence for all persons. . . . We have pondered upon these words. Our contemplation leads us to state that the Church in its struggles to be faithful to its call for reverence and dignity for all persons must respond by providing the possibility of women as persons being included in all ministries of our Church.*"

Loud applause broke out. When John Paul II moved to the podium, sisters wearing blue armbands—the symbol of women's ordination—rose one by one, until some forty-five were standing in silent protest. They remained standing until the ceremony ended.

Kane's stand emboldened a movement. Women began in the 1970s to flock to the nation's theology schools to earn the master's in divinity—the degree held by ordained priests—for the first time in American history. They were certain they were preparing for ordination to the Roman Catholic priesthood in their lifetimes.

In 1981, the Jesuit-founded Center of Concern in Washington sponsored the Women Moving Church conference, organized by then Sister Diann Neu and attended by laywomen, nuns, and clerics. It moved beyond the question of women's ordination to other issues affecting women in the Church, from the persistence of sexist liturgical language to a call for "a renewed emphasis on cooperation, solidarity, mutuality, shared responsibility and decision-making, and participatory structures."

At that conference, the subject began to shift from Catholic women impacting the Church to Catholic women recognizing themselves *as* the Church. Referring to the long history of women organizing into religious communities in the Catholic Church, theologian Elisabeth Schüssler Fiorenza told the crowd: "A Catholic feminist spirituality claims these communities of women and their history as our own heritage and seeks to transform them into the *ecclesia* of women by claiming our own spiritual powers and gifts." The word *ecclesia* is Greek for "church," defined as the actual gathering of people.

Linguistically, it was a short trip from an "*ecclesia* of women" to Women-Church. That was the name given to the nationwide network of women's worship groups—modeled after Latin America's Catholic small-base communities—that began to form in the 1970s. From the early 1980s through the mid-1990s, more than five thousand women active or interested in such groups attended major conferences mounted first by Women of the Church Coalition and then by its successor, Women-Church Convergence.

Some of these groups celebrated the Eucharist themselves, as women had done at the second Women's Ordination Conference. But others created their own liturgies, seeking to mark the milestones in their lives, milestones ignored in traditional liturgical practices or denigrated. In response to a Church that once banned women soiled by childbirth from entering their parishes until they were purified by a "churching" ritual, for example, these women reformers brought women's bodies and blood back into the realm of the sacred. Feminist liturgies abounded in the decades that followed, from "Honoring Women's Blood Mysteries" to "A Ritual of Healing from Childhood Sexual Abuse."

When the Bishops Listened

The pressing nature of women's issues post-Vatican II led the bishops to vote unanimously in 1983 to develop a pastoral letter on women's concerns about the Church and society. A committee of six bishops and seven female consultants authorized "listening sessions" with some seventy-five thousand women across the country, including members of the Women's Ordination Conference and other women's organizations.

The first and most popular draft with the progressive Church reform groups was entitled *Partners in the Mystery of Redemption*, published in 1988. In it, women—often quoted in their own words—shared their concerns in great emotional detail over numerous issues involving the Church and the world in which they lived. They wanted expanded opportunities for ministry, the chance to be ordained, support for balancing home and family, help for domestic violence and sexual abuse, and a revamping of the Church's teachings on sexuality, especially the ban on artificial birth control.

In that version of the document, the bishops named "sexism" a sin. They vowed to work for women's rights—from better access to child support to the active participation of men in child-rearing to equal pay for equal work. They conveyed women's "voices of affirmation" as well as their "voices of alienation," quoting women sympathetically in a way that would be unthinkable in Church documents today. "How can the hierarchical church, which claims to be in search of the truth, feel reasonably comfortable that it has found the truth when in decisions regarding procreation . . . the decision-makers represent only the male half of the human perspective?" read one woman's remarks.

The pastoral went through four published drafts over ten years. Critics weighed in. The very form of the enterprise incensed many feminists, who felt that the bishops were treating women as the problem. "When they did their pastoral on racism, they didn't do it on black people," explained Women's Ordination Conference organizer Dolly Pomerleau. "And when they did a pastoral on the economy, they didn't do it on poor people." The subject, those women felt, should not have been women, but sexism.

Vatican officials reviewed each draft, moving the language further and further away from a spirit of dialogue to one of laying down the law. The fourth version of the pastoral was published in 1992. Women's quotes were gone and with them women's voices. The tone had become increasingly paternalistic. And the Church's teachings were presented as fixed. The bishops could not agree to approve the document. The pastoral failed to pass, an outcome unique in the history of the U.S. bishops' conference.

A Legacy of Backlash

As the post-Vatican II movement for change evolved, its goals became increasingly dangerous to an increasingly conservative Catholic Church hierarchy. Once again, the iron fist of Vatican authorities slammed down on women in an escalating campaign to silence the voices for change.

The first major setback came in 1968 from Pope John XXIII's successor, Paul VI. He took over the special Papal Birth Control Commission John XXIII had established to deal with that most contentious issue. Hailed as an essential step to closing the growing rift between the people in the pews and the Church's ban on all forms of artificial birth control, the commission had representation from bishops, theologians, scientists, demographers, and several married couples. While hopes were high that change would come, Paul VI shocked the world by completely defying the majority recommendation of his commission to end the birth control ban. Instead, he unequivocally condemned artificial contraception in his now infamous encyclical *Humanae Vitae*. That left the Church still approving only "natural" family planning—the rhythm method, which calls for abstinence during the woman's fertile times—thereby green-lighting the intention to contracept but arbitrarily limiting it to only one means.

A few years later, in 1976, Pope Paul VI rejected the findings of yet another commission, the Pontifical Biblical Commission on the subject of women's ordination. The majority membership of that commission had found insufficient scriptural grounds to exclude the possibility of women's ordination. That same year, the Vatican Congregation for the Doctrine of the Faith issued the notorious encyclical *Inter Insigniores*, forbidding women priests. That document argued that women could not be ordained because, being female, they could not "image" Jesus. On the basis of the symbol of Christ as the bridegroom and the Church as his bride (never mind that laymen as well as laywomen play the bride's role), it further maintained that priests—in Jesus' stead—must therefore be men. The encyclical chastised those who would measure the Church by standards of equality that were being imposed on all other institutional structures, saying that "the priestly office cannot become the goal of social advancement." Rather than recognize the sincerity of a woman's call to the priesthood, it blithely concluded that "such an attraction, however noble and understandable, still does not suffice for a genuine vocation" but rather represents "a mere personal attraction." And the document clung to its imagined ancient tradition of an all-male priesthood, employing a pattern of minimizing, denying, and completely ignoring what the Bible actually says about the roles of women in early Christian communities.

But the greatest force against the rise of women change-makers in the Catholic Church in the aftermath of Vatican II has been Pope John Paul II, whose reign began in 1978, making it the third longest in papal history. John Paul II immediately set about replacing bishops sympathetic to women's and other progressive Church issues with reactionary recruits, increasingly committed to turning the clock back to the days before Vatican II. Eventually, John Paul II would appoint more cardinals than any other pope, all but five of the world's 135 cardinals under the age of eighty by whom the next pope will be chosen.

John Paul II sees himself as an advocate for women and women's equality. It's true that during his reign, women have risen to high-level positions—for instance, as chancellors and vicars—in dioceses nationwide, and to positions never held before at the Vatican, such the first women members of the International Theological Commission. And he has energetically championed women who work in "social, economic, cultural, artistic, and political life." But at the same time, he has kept women locked out of the highest levels of authority, holiness, and sacramentality within the Church. While denouncing discrimination against women, John Paul II made only a halfhearted apology for the Church's terrible role in that discrimination, offering it only "if objective blame" could be found. He has written of his commitment to "upholding the dignity, role, and rights of women" but has never budged from Church teaching on complementarity of the sexes, which relegates women to a subservient, service-based role. "Ordinary women . . . reveal the gift of their womanhood by placing themselves at the service of others in their everyday lives," he has written. "For in giving themselves to others each day, women fulfill their deepest vocation." In an indication of what enigma women are to him, John Paul II talks about the "mystery of women," the "genius of women," and God's "mysterious plan regarding the vocation and mission of women in the world," language that would never be applied to men.

John Paul II has been unrelenting in his determination to keep women out of the ordained priesthood. In 1994, he issued a scaldingly negative apostolic letter, *Ordinatio Sacerdotalis*, "On Reserving Priestly Ordination to Men Alone," which upped the ante on Church teaching. Deriding those who considered the matter of women's ordination "still open to debate" and who saw the Church's ban as having "merely disciplinary force," the pope decreed that "the Church has no authority whatsoever to confer priestly ordination on women," and "this judgment is to be definitively held by all the Church's faithful." It was in effect a silencing order; it declared the case closed. An uproar from Catholic women around the country ensued, which did nothing to soften Vatican language but instead inspired an even more thunderous clerical crackdown. The next year, Cardinal Joseph Ratzinger, the prefect for the Vatican Congregation for the Doctrine of the Faith, went a step further than Pope John Paul II, declaring the teaching against women's ordination to be "infallible." A few years later, the Vatican expanded canon law so that Catholics who refused to accept certain Church teachings—including the Church's refusal to ordain women—could be excommunicated.

John Paul II has set in veritable stone the Church's opposition to birth control, sterilization, infertility treatments, condoms even to prevent the spread of AIDS, the morning-after pill even in cases of rape, stem cell research, and pregnancy termination in virtually all circumstances. Particularly through the 1990s, the Vatican played a major obstructionist role on matters relating to family planning at the UN women's meetings. John Paul II has consistently condemned homosexual behavior in the harshest terms, as well as divorce and remarriage.

And for many years, the Vatican has labored to decimate initiatives—including those approved by American bishops—to adopt inclusive language for God and for humankind in the Latin-to-English translations of the missal (Mass prayers), the lectionary (Bible readings read at Mass), Scripture (Old Testament, including the Psalms, and the New Testament), and the Catholic catechism—activity sparked by Vatican II's advocacy of Church language in the vernacular. The opposition of the Vatican to inclusive language has been vocal and virulent. In the latest version of the Catholic catechism, for example, the Vatican refused to accept even the most unimposing inclusive language, replacing the proposed "for us and for our salvation" in the Apostles' Creed with "for us men and for our salvation"— just one of numerous similar examples. A particularly powerful body of eleven men appointed by the Vatican—whose identities were kept secret until uncovered by the *National Catholic Reporter*'s John Allen—determined the final language for the Bible passages in the American lectionary, throwing out earlier inclusive versions approved by the U.S. bishops. Asked why there was no woman on the committee, lay member Michael Waldstein, a university professor and head of a conservative theological institute in Austria, was quoted in the *National Catholic Reporter* as saying: "The issues are well known. I don't feel having a woman present would have added anything."

Though less obvious, the Church's persistent patronizing and negative views of women have influenced other reactionary stances as well. Its views are evident in the Church's historically derisive and dismissive treatment of the mothers of children who have been abused by priests. Those views influence the Church's rigid opposition to married Catholic priests, which precludes women having an approved intimate connection to the Church's inner clerical circle. And those views are apparent in the Church's willingness to use the labor of women to assume a vast array of ministerial responsibilities, without giving women the ultimate spiritual recognition, sacramental authority, administrative power, or doctrinal influence.

During this pope's tenure, name-calling and scapegoating have thrived at the local level, too. A trail of bitter invectives against female reformers has appeared in the *Southern Nebraska Register*, the paper of the Lincoln diocese, headed by the ultrareactionary Bishop Fabian Bruskewitz. In the *Register* Sister Jeannine Gramick—who began decades ago to minister to Catholic homosexuals—was dubbed an "apostle of sexual perversion." Then eighty-four-year-old Patty Crowley, a mother of five who served on Pope Paul VI's Birth Control Commission, was labeled a "very old degenerate who roams about promoting sexual immorality." Preeminent Catholic theologian Rosemary Radford Ruether was called an advocate of "witchcraft," and Sister Joan Chittister "a pope-hating feminist." In 1996, Bruskewitz threatened members of Call to Action, Catholics for a Free Choice, and Planned Parenthood with excommunication. Increasingly through the years, many dioceses and parishes have banned CTA members—as well as members of the Women's Ordination Conference, the Catholic homosexual rights group Dignity, and other post-Vatican II reform groups—from meeting on church property or speaking at parish or diocesan events.

Right-wing books and websites attack women reformers, too. "Dominican Retreat House Hosts Lesbians and Witch" was the headline that appeared on the right-wing website *Les Femmes* in 2000. It referred to the founders of WATER (Women's Alliance for Theology, Ethics, and Ritual)—Diann Neu, the former nun who organized the Center of Concern's 1981 Women Moving Church conference, and her partner, highly respected theologian Mary E. Hunt—as well as author and artist Mary Lou Sleevi. All three were to participate in a women's spirituality series to take place in McLean, Virginia. While Neu and Hunt are out lesbians, publicly active in Church reform, Sleevi is a married woman with apparently closer ties to the institutional Church. She told the press that her artwork has been shown several times at meetings sponsored by the U.S. bishops.

In response to the complaints, Bishop Paul S. Loverde of the diocese of Arlington, Virginia, canceled the series. Demeaning Neu and Hunt's organization, WATER, he wrote in an open letter: "If the water in the well is allowed to become polluted, no one should be surprised when the people who drink it become ill." He slammed Sleevi, too, objecting to the association of her paintings with "words of empowerment." Such words "sound innocent enough," he wrote, "but . . . take on a whole new meaning when they are applied in a feminist context." True to his convictions, Loverde also forbids female altar servers throughout his diocese.

Embattled Movement Holds On

Progressive reformers suffered from being demonized and marginalized by the Church hierarchy during the reign of John Paul II, and many lost hope. But the reform organizations held on, providing places of refuge for Catholics struggling to keep the flame of Vatican II alive.

By the turn of the twenty-first century, women were playing a dominant role in America's progressive Church reform community, leading and coleading all of the major organizations. Call to Action's membership had reached twenty-five thousand, making it the nation's largest Church reform group; women represented its largest constituency, its major spokespeople, and all of its presidents. The Women's Ordination Conference had weathered financial woes, administrative upheaval, and an internal struggle between the Women-Church forces committed to independent ministry and those fighting for ordination within the Church, accepting both; it went on to help found Women's Ordination Worldwide. Links grew stronger between the American and international Church reform communities, with an American nun and a former priest becoming the U.S. representatives to International Movement We Are Church, which succeeded in gathering 2.5 million signatures on a petition to the Vatican supporting a broad range of Church reforms.

And women reformers had begun to move full force in new directions. To the dismay of the hierarchy, they began to publicize the crippling priest shortage. They promoted awareness of women's crucial roles in the ancient and the contemporary Church. One survivor of priest child sexual abuse single-handedly mounted a national campaign that became a voice for victims nationwide. Reformers intensified their challenges to the Vatican's opposition to issues like condom use to prevent the spread of AIDS. More and more Catholic feminist theologians published their provocative views on everything from sexual ethics to the face of God. Catholic women moved into an ever wider array of ministries. And the actions of towering women reformers—such as Sister Joan Chittister in her stand against silencing, Sister Jeannine Gramick in her defense of Catholic homosexuals, and wife and mother Mary Ramerman in her struggle for ordination—had become increasingly bold.

By the time the priest pedophile crisis broke, the progressive Church reform movement in America had begun to rumble again, a distant rumble, like rising thunder before a storm.

3

✠

NEW REFORMERS, OLD REFORMERS FACE OFF

Less than a year after the Erie Benedictines took their stand, their courage became an inspirational backdrop for the event that sent the Church reform movement to exploding: the pedophile scandals of 2002. An avalanche of secret Boston archdiocesan files, released by court order, documented a horrific history of Catholic priests' serially raping and sexually molesting hundreds of minors, beginning with a notorious pedophile Father John Geoghan. Instead of being brought to justice, Geoghan and others like him were hidden within an institutionally sanctioned web of cover-ups and lies, spun by Cardinal Bernard Law of the archdiocese of Boston with other of the nation's most prestigious bishops and cardinals. In a series of Pulitzer Prize–winning stories, the *Boston Globe* set off a firestorm of similar revelations by news reporters in dioceses around the country, which created for American Catholics a searing and devastating moment of truth.

Loyal parishioners at Saint John the Evangelist Church in Wellesley, Massachusetts, faced such a moment. Outraged by what they had learned, some forty of them, with the support of their devoted pastor Father

Thomas Powers, took to the microphone after Mass on the second Sunday in February 2002, to share their disillusionment and pain. Out of those early meetings came a flourishing national movement that would soon eclipse Call to Action as the largest Church reform group in America: Voice of the Faithful (VOTF). From its inception, VOTF identified itself as a centrist organization and refused to be associated with the stalwart Vatican II progressive reformers and their agenda. VOTF's public voice was decidedly male, and women's issues were decidedly buried.

A New Voice for Change

There is no question but that the pedophilia scandal—at least temporarily—tore the mantle of sanctity off the Catholic Church. Long subject of a litany of illogical and unconscionable sexual prohibitions, Catholics were forced to see how the hierarchy had failed to live up to even the most basic moral standards. That, explained William D'Antonio, Catholic University's sociologist at the time, exacerbated their outrage. His research showed that only 20 percent of American Catholics looked to Church leaders as a source of moral authority on matters from birth control to abortion, homosexuality to nonmarital sex to remarriage after divorce, and that was *before* the scandals broke. Most Catholics also believed that the Church should be more democratic. What our data shows, D'Antonio told me, was that people are "ready for the reform movement."

And the reform movement was ready for them. But the influx of newcomers posed a real challenge. Catholics from all over the political spectrum—many never active before in Church affairs—demanded a voice. Some joined the old-guard progressive reformers, while others were mapping out a new middle ground. What was different, observed Tom Fox, publisher of the *National Catholic Reporter*, was the emergence of "a large cross-section of Catholics calling for change. From left to right, Catholics are calling upon the bishops [for] greater accountability," he told me. "They're saying the Church that we have, the governance we have, is not working."

In addition to holding clericalism and secrecy responsible for the crimes, the old-guard reformers saw in the crisis dramatic evidence of the need for change in their most contentious areas of concern. Advocates of gay rights in the Church, they were incensed by the hierarchy's shameless scapegoating of homosexual priests as the cause of the scandals. In fact, they pointed out, that contention could not be separated from the Church's denial of the rampant sexual abuse and exploitation of girls and women by Catholic priests, their conclusion being that neither homosexuality nor heterosexuality per se were causes of the sexual abuse of children. They also saw women's overall subordinate role in the Church—including the ban on ordination—as a major cause of the crisis. Call to Action spokeswoman Linda Pieczynski reflected that general sentiment when she said that if women—particularly mothers—had been at the table, "We wouldn't have tried to save Father Bob's reputation. We would have protected the children."

By contrast, the leaders of the new "center" of the reform movement chose not to clutter their message with a litany of "hot-button" issues—for or against women's ordination or optional celibacy or gay priests—in an effort to join all Catholics, from the left, right, moderate, traditionalist, and progressive camps, into one voice for change. Their task was enormous, but their slogan was simple: "Keep the faith; change the Church."

VOTF's main concern was not that the laity was unable to influence particular issues in the Church but that lay Catholics had no voice at all. VOTF demanded that the Church hierarchy begin to share power and responsibility with the people in the pews, and they based their demand squarely on Church doctrine. They referred to the *sensus fidelium*, the required sense of the lay faithful on Church doctrine and beliefs; to Vatican II teachings that the laity had not only a "right" but "a responsibility . . . to become active in the guidance of the Church"; and to canon law, specifically Canon 215, empowering Catholics "to found and govern" lay organizations.

Challenging the "Father knows best" mode of Church leadership, VOTF sought accountability, transparency, and an end to the secrecy and closed clericalism that enabled the sex abuse crisis and cover-up to occur. VOTF swiftly moved forward with a specific three-point agenda: to support clergy sex abuse survivors, to support "priests of integrity," and to "shape structural change" in the Church. The most threatening item on VOTF's agenda, structural change, was eventually defined as ensuring that the laity have a real voice in Church-approved governing bodies, such as safety, pastoral, and finance committees.

In short order, VOTF built an expanding network of independent, parish-based groups all over the country. VOTF declared its members to be the very backbone of the Church, drawn from the ranks of lectors, Eucharistic ministers, religion teachers, deacons, choir members, and others active in parish ministries. VOTF set up a charitable fund through which Catholics could redirect their donations from Boston archdiocesan coffers to Catholic agencies and became its own nonprofit entity. And VOTF held its first conference—a national one—just months after its formation, in the summer of 2002, which drew an emboldened crowd of four thousand.

While VOTF has members who also have participated in the old-guard reform groups, the organization from the start did not consider itself part of the progressive community. Cofounder James Muller, a cardiologist who also cofounded the Nobel Prize–winning International Physicians for the Prevention of Nuclear War, described VOTF as a centrist organization early on. "We are building a representative structure for the laity that will be more like Congress than the Democratic or Republican Party," he wrote to me. At the same time, VOTF went to great lengths to make conservatives feel welcome while distancing itself from progressive groups. It was also careful not to take positions on what one reformer called "issues of controversy," like women's ordination.

Church Takes on VOTF

Remaining separate from the old-guard progressive reform community in the eyes of the institutional Church proved a challenge for VOTF. Even in the grips of the scandal, the institutional Church continued to demonize Church reformers, insisting that their dissent had created the culture of permissiveness that led to the sex abuse scandals. Faced with a range of reform groups—from Call to Action to Dignity to the Survivors' Network of those Abused by Priests to Voice of the Faithful, all clamoring for change—the besieged hierarchy didn't bother to make distinctions.

Under mounting pressure from reform groups, however, the Catholic hierarchy was forced to come out of the clerical closet and respond to the public outrage over clergy child sexual abuse. Beginning by devoting its general meeting in Dallas, Texas, in June 2002, to the crisis and opening that meeting to the media, the United States Conference of Catholic Bishops met the crisis in a way it had never done before. They invited victims and influential lay Catholics outraged by the crisis to publicly confront the entire assemblage of some three hundred bishops. They drew up norms for sex abuse policies and committed themselves to "zero tolerance" by pledging to permanently remove any priest from the ministry found to have abused a minor child.

The U.S. Bishops' Conference created a *Charter on the Protection of Children and Young People*—setting out policies for the dioceses in handling allegations, dismissing abusive priests, reporting to civil authorities, reaching out to survivors, and preventing future abuse through Safe Environment programs. They established an Office of Child and Youth Protection to monitor the application of the charter's principles and policies, and hired as its executive director Kathleen McChesney, who had been the FBI's third-highest official and its highest-ranking woman ever. The conference committed to auditing compliance with the charter in the form of an annual public report, to be prepared by McChesney's office.

The conference also appointed thirteen high-powered Catholics—from former President Bill Clinton's chief of staff Leon Panetta to former Oklahoma governor Frank Keating to Chicago appellate court Justice Anne Burke to former federal prosecutor and special counsel to the U.S. Senate Ethics Committee Robert S. Bennett—to serve on a National Lay Review Board. Conceived from the start as an independent body, the board was charged with overseeing the work of the Office of Child and Youth Protection, including the annual compliance audit; commissioning a study into the extent of child sex abuse by Catholic priests; and conducting its own nonscientific survey into causes and context of the crisis.

The relinquishment of even some hierarchical authority did not sit well with many of the prelates. In his closing remarks to the bishops at their Dallas meeting, Chicago's Cardinal Francis George held the Church's opponents—a list headed by women and feminists—responsible for the hierarchy's diminishing power. George's targets included the campaign of

Catholics for a Free Choice (CFFC) to downgrade the Church's status at the United Nations to a nongovernmental organization; feminists working for laws requiring Catholic institutions that serve and employ the general public to provide reproductive health services, including contraceptive insurance; anyone suing the Church; Catholics with a shaky faith; Protestants; and American culture in general.

He came just short of calling the whole lot a "cabal." Reflecting similar sentiments, America's cardinals, upon their return from a high-profile visit to Pope John Paul II in April 2002 to discuss the breaking crisis, issued a statement instructing pastors "clearly to promote the correct moral teaching" and "publicly to reprimand individuals who spread dissent."

Illustrating the extent of the hierarchy's paranoia, in 2002, Cardinal Law forbade an effort by existing diocesan lay councils under the Church's jurisdiction—not even part of the Voice of the Faithful movement—to form a simple association. After many rebuffs, VOTF representatives were able to meet with archdiocesan representatives, an unfriendly meeting at which Cardinal Law's top aide, Bishop Walter Edyvean, reportedly invited them several times "to shut your group down." Early on, Law banned VOTF groups from meeting on Church property—a ban long imposed on old-guard reformers. Within two years, eight dioceses nationwide and individual parishes joined in that VOTF ban.

Despite their success at avoiding support for the "hot-button" issues, VOTF members were soon declared to be heretics and dissidents and anti-Catholic by conservative forces in parishes, dioceses, and the press. At the same time, some victims and their supporters considered VOTF far too compliant and friendly with the bishops as well as insufficiently focused on victims' issues; they left to form other organizations, like the Coalition of Catholics and Survivors and Survivors' First.

But VOTF never wavered. The organization continued to grow in numbers and power. While at first unwilling to do so, by the end of 2002, VOTF called publicly for Cardinal Law's resignation. So did members of the new Priests' Forum, another group formed in the wake of the sex abuse scandals, which works closely with VOTF. Cofounder Father Walter Cuenin, a Newton, Massachusetts, parish priest, helped lead the drive to collect fifty-eight priests' signatures on a petition calling for Law's resignation because of his failure as a spiritual leader—an action that expert observer and theologian Richard P. McBrien called unprecedented in modern American history. Cuenin had earlier named the Erie Benedictines as one of his influences. What they did, he told me, "had an impact on me personally. A lot of us have lived in fear—I can't speak because something will happen. If enough people speak, there's nothing that anyone can do." In Law's case, enough people spoke; in December 2002, the pope accepted his resignation.

The Women of Voice of the Faithful

Despite positioning itself at the center, refusing to take a public stance on such issues as women's ordination, and its public male voice in the person of Dr. James Muller, VOTF is full of women leaders who play critical roles in the organization's work. What I discovered was that beneath VOTF's surface of neutrality about women's issues is a great force for change in women's status in the Church that could well bubble up in time. In fact, as I observed the women leaders, that force began to bubble up before my eyes.

Significantly enough, a woman, Susan Troy, is VOTF's spiritual director. A founding member of VOTF who attended that first meeting at St. John's parish, Troy, fifty-six, a wife and mother, holds a master's of divinity from the Weston Jesuit School of Theology in Cambridge, Massachusetts. Troy is also an independent spiritual director by profession, offering today's lay Catholics the comfort of "spiritual companionship," based on the model of spiritual guidance that Catholic priests have provided to monks for centuries.

To Troy, VOTF is not a reform organization but a faith movement. She explains that the initial force for the creation of the organization was not reform, though reform may result. "People's faith was decimated. . . . They were lost," she says of the beginnings of the sex abuse crisis. "This entire movement came out of people's faith. It's an articulation of their spirituality, how you put your faith system to work in the world."

Troy is among the many Catholic women who have felt called to be a priest, which led her to study at Weston. She estimates that about a quarter of her 1998 graduating class were women; the rest, Catholic seminarians. The experience, she confesses, was "excruciating." Describing what it felt like to earn a master's in divinity and then be forced to sit on the sidelines as her male classmates were ordained—priests one day, deacons the next—she said: "It's like, if you finish medical school . . . everyone's cheering for you, is so excited. You got A's and you're smart, and you've been having great discussions and proving your skills, and professors have been wowed by you. Then you all graduate, and they say: 'I'm sorry, only people with penises can be doctors. The rest of you, not [registered] nurses, but LPNs.'" She estimates that the majority of the Catholic laywomen in her class have left the Catholic faith. She and the few other women who have stayed keep in touch, providing one another support.

When asked honestly if she ever thinks about leaving the Church, Troy says with a laugh, "Oh, God, on a daily basis." Then what keeps her? "This is my faith," says Troy, an Irish Catholic, of the Church into which she was born. She describes herself as "terribly sacramental," a believer that sacramentality is what the Catholic Church offers more than any other Christian denomination, and a lover of the Eucharist, which, she says, "is it for me." But there is another reason. "I went ahead in the Catholic Church as a woman to make my voice heard," she told me. "The reason I don't leave is because I feel a solidarity with all women in the Catholic Church. . . . I have a privileged position, by dint of my struggles and my education, and I don't want to leave them voiceless." When asked what she wants Catholic women to do with that voice, she thinks a moment, then says: "I want them to take back their Church."

Troy may not be a priest, but she has great confidence in her Church knowledge and her spiritual authority. Were it not for Troy, there would have been no Mass at VOTF's first national conference in 2002. Concerned that Cardinal Law's permission was needed and he wouldn't give it, priests involved with VOTF recommended a simple non-Mass liturgy. Troy argued fiercely that VOTF could not bring together four thousand Catholics and have no Mass. She insisted that having the Mass at the convention site would not violate Church regulations, as the priests feared, and that Cardinal Law's approval was unnecessary. She refused to relent, the priests finally agreed, and the Mass was held. It represented a "coming-of-age moment," said Troy. "The Eucharist belongs to all of us. It's not one of the assets of the Roman Catholic archdiocese of Boston."

Peggie Thorp—wife, mother, and writer, fifty-five—is another founding member of VOTF. After that first meeting at St. John's Church, Muller asked Thorp to join him in cofacilitating future meetings, and she agreed. The Monday night gatherings that followed built and formed the organization. That job was a real challenge for Thorp, who had long ago become disillusioned with the Church. "I sort of dismissed the entire institutional Church when I was very young, in my middle-twenties," she reports. "But I still clung to my faith, and I loved liturgy and the Mass. Still," she says, "It was always very hard to be reminded in such a face-to-face way every Sunday that you are a second-class citizen in the institution that has, for better or worse, preserved the faith that you depend on for two thousand years."

Thorp raised her children in the Church and taught religious education classes for a time but confesses to having "had no interest whatsoever in the politics of the institutional Church." In fact, she told me, before the scandals broke, she had been doing research for a book intended to explore why "any intelligent, progressive-minded Catholic woman" would ever "stay in this misogynistic culture." She planned to use what she learned to advise other women about how they, too, might stay in the Church. The prescription she arrived at: Keep the institution at arm's length. It is a prescription that for her, the sex abuse crisis completely "upended." Now she sees involvement as the only way to stay.

Thorp and Muller cofacilitated the meetings attended by an ever-growing cadre of Catholics from all sides of the progressive-conservative divide, all willing to commit to VOTF's three moderate goals and leave their other issues outside the church doors. That took discipline from Thorp, and she learned a lot from that experience. "As a facilitator, I had to do three months of listening," she explains. "Your job is to get other people to talk and feel comfortable doing so. . . . [L]istening to these people I would never have listened to before, it was really life-altering. All these people with all these different perceptions and politics are just like me. We all love the same Church."

Muller then invited Thorp to become VOTF's first vice president but she declined, wanting instead to be VOTF's written voice, penning the organization's growing mountain of paper and web literature. She's devoted to the work and loves the organization but admits that her personal faith has been shattered by the crisis—not in God, but in the hierarchy. "It would be wonderful to go back to church and have things the way they were and really believe in the authority of the priests and the bishop," she says. "But I could never do that again because I've finally come to a point where I don't recognize their authority because they don't recognize mine. And I'm not afraid to say that anymore."

As to why Thorp stays in the Church, her response is impassioned. "Because I wouldn't give them the satisfaction of leaving," she says. Her work for VOTF is key. "I'm in there to try and do what I can to move more Catholics . . . to expressing themselves, to voicefulness. . . . I'm in there so that if I ever do leave, I will have reached as many Catholics as I can. . . . They have a right to speak and a right to be heard. After that, they have a right to follow the Christ they recognize."

Svea Fraser, also fifty-six, is another cofounder of VOTF as well as a wife, mother, holder of a master's in divinity, and the first Catholic female chaplain at Wellesley College. She left that job to devote her energies full time to VOTF, where she now heads the Priests of Integrity Working Group and serves on VOTF's board.

While outraged by the scandals and the pain they have caused survivors and their families, she also has particular sympathy for the innocent priests impugned by the scandals and even for former Boston Cardinal Bernard Law. "I have always seen Cardinal Law as a victim, too," she says, though not to the "painful extent" she sees survivors as victims. "He was a company man, and he did it the company way." To her mind, the crisis is "about a system, not about a person. It's a feudal system, with no transparency and no accountability [that] allows for the terrible abuse of power." She sees VOTF as "a prayerful voice" through which the faithful—by which she means all Catholics, cleric and lay—can work together, breaking down barriers within the Church, especially between the episcopal leaders and the people.

While men still served as president and executive director of VOTF in 2004, nearly all other national staff were women. More than half of VOTF's regional coordinators and a third of its affiliate leaders were women. Founder and coordinator of Brooklyn-Queens regional VOTF, thirty-seven-year-old Melissa Gradel is a mother, wife, and arts administrator. She returned to her Catholicism as a young adult to the great dismay of her father, who had left the Church years before. "May you be the kind of Catholic bishops live in fear of," he advised. Soon, she was. In January 2003, as the leader of a VOTF delegation, Gradel faced off with former Bishop Thomas V. Daily's right-hand men, lobbying—successfully—for the rescinding of the Brooklyn archdiocese's ban on VOTF's meeting on church property.

Enraged by the sex abuse scandal and women's complete lack of power and voice in the Church, Sandy Simonson, thirty-two-year-old mother of two, launched a fierce letter-writing campaign to then Phoenix Bishop Thomas O'Brien. In response to her personal letter saying that she planned to withhold donations until the silence about sex abuse ended, and that women needed to play a larger role—including being ordained—for the Church to move past the crisis, O'Brien wrote an amazingly arrogant reply. He accused her of having her own "agenda" and joining in the media "frenzy." He said that her withholding donations would not affect "the diocese or me in any particular way." And in closing, he wrote: "Your letter indicates that you do not understand your role as a Catholic in the Church or the way that leadership is to operate."

Actually, Simonson did understand. Soon after, she founded and headed up the Phoenix area VOTF. Thus empowered, she became an outspoken VOTF advocate and wrote another letter to O'Brien, this one on VOTF letterhead, publicly urging him to resign. In the end, he did. O'Brien avoided criminal indictment on obstruction of justice charges by admitting that he had knowingly transferred abusive priests to other assignments without informing their superiors of their crimes and by agreeing to turn over control of some diocesan matters—including the handling of all sex abuse allegations—to an independent authority. Even then, the pope did not demand O'Brien's resignation. But shortly thereafter, in June 2003, O'Brien was the driver in a fatal hit-and-run accident, after which he finally did resign, and for which he was eventually convicted.

More women than men continue to be nominated for the presidency of

VOTF, but they withdraw their names before the voting starts. Some do so because they are too busy. But there is another reason at work. Svea Fraser has been nominated for two years running but withdrew her name both years. "We still struggle with the fact that we know that having a man for our president was going to have more clout in the world because that's the way it is," she says. She describes the situation as unfortunate and says the hope is that the next president will be a woman.

As for VOTF's foot soldiers, women appear to predominate at many parish gatherings, chapter events, and regional conferences, like VOTF's New York area regional conference, Being Catholic in the Twenty-first Century, held at Fordham University in 2003 and attended by some 1,500 people. As in the old-guard groups—to which some VOTF members also belong—those who participate in VOTF's activities tend to be middle-aged and older. And like the old-guard groups, VOTF must work hard to recruit younger members—illustrated by the fact that all four young women who served on the "Voices of the Future" panel at the Fordham conference had never even heard of VOTF. All of them expressed support for the organization, once they learned what it was, despite the fact that they came from very different vantage points. One young woman defended the Church's slow pace of change on subjects like women's ordination, pointing out that the United States was nowhere near ready to elect a female president, while another berated the institution from which she felt alienated for "living in the Ice Age."

Change-makers Face Off

During Easter Week in 2004, Archbishop Sean P. O'Malley, Cardinal Law's successor, dashed the hopes of many women for healing in the diocese most decimated by the sex abuse scandals, the diocese in VOTF's backyard. On Holy Thursday, in a reenactment of the ancient ritual of Jesus' washing his disciples' feet, O'Malley invited homeless men, male parishioners, and several religious brothers up to the altar at Holy Cross Cathedral but refused to wash any woman's feet. The same thing happened in Atlanta, where Archbishop John F. Donoghue banned all priests from allowing women to participate in the foot-washing ceremony, which led to protests organized by

members of the Women's Ordination Conference and some parishes cancel-
ing services altogether. Adding insult to injury, in his Easter Week Mass for
hundreds of priests, O'Malley listed "feminism" among the world's worst ills,
lumping it along with "the drug culture," "hedonism," "consumerism," and
"the culture of death." In a throwaway line, he also referred to older Catholic
women as "biddies."

As fate would have it, just a week or so after O'Malley's assault on
Catholic women, more than five hundred—mostly women—attended
Boston College's sold-out weekend conference devoted to "Envisioning the
Church Women Want," part of its *Church in the Twenty-first Century* series.
Many women arrived outraged by what had just occurred.

The keynote speakers were female reformers with long histories in the
old-guard progressive reform movement. In the audience also were the national leaders of VOTF: Peggie Thorp, Susan Troy, and Svea Fraser. It didn't
take long for the sparks to fly. Ironically, some of the "moderate" new reformers found the old-guard reformers to be insufficiently radical.

Beloved theologian Ada María Isasi-Díaz, a leader for Church reform
for more than thirty years, delivered the opening keynote. In an address entitled "The People of God's Church in the Twenty-first Century," Isasi-Díaz
argued for replacing the "one, holy, Catholic, and Apostolic Church" with
new markers. She called instead for a Church that is "humble"—that is,
which faces "messy reality" head-on; "pluralistic," that is, which sees diversity
as a blessing, not a problem; and "prophetic," that is, a Church of the poor
and oppressed that trusts desire instead of suppressing it and sees desire as
the fuel for revolution. She argued, as she has argued for decades, that we are
all—not just the clerical ranks—the Church, and that, with or without the
clerics, ordinary Catholics can themselves model this Vatican II a "humble,
pluralistic, and prophetic" Church.

But then all the frustration burst forth. "Humble?" asked an incredulous
woman who rose to speak. "I feel like I've been humble long enough." Another said, "We've been 'envisioning' for thirty years. Are we any better off?"
Around the room the question loomed: What should we *do*? Isasi-Díaz responded, facetiously: "All we can do is hope and pray and fast because some
spirits are hard to drive out." But then she added: "Polity is from the top

down. It's impossible to change the polity of the Church. You need another John XXIII," the convener of Vatican II.

"If we are the Church, how can we say that the institution will never change?" VOTF's Peggie Thorp insisted to me later, deeply dismayed by Isasi-Díaz's remarks. Besides, she added: "We've *already* changed the institution. We've changed it. It will never be the same because the laity will never again be silent." Of the clerical leaders, she said, "We have hauled them into the arena of discussion. . . . They are required to respond. They can't work the way they used to work. . . . I don't think they've had a sleepful night since January of 2002. And if that's all we do for the rest of my life, I'm happy."

Susan Troy sees VOTF's progress in the fact that there is now "a unified lay voice. Even though it's sometimes only manifested in the press, at least [there is] a growing mind-set that if we hear from the bishop, there's somebody else that we need to hear from, too." In that sense, she believes, "we're changing the conversation."

VOTF leaders Thorp, Troy, and Fraser were upset that no survivor of clergy sex abuse was scheduled to deliver a keynote or lead a workshop, nor any "ordinary Jane Q. Catholic," like a member of VOTF. Troy stood before the whole conference audience and issued a protest to that effect, at the close of Isasi-Díaz's presentation. The VOTF leaders also were unhappy about the lack of emphasis on action. Thorp wants to see people doing something, whatever they feel they can do, even if that is only standing in protest at a Mass where they cannot abide what is being said. "Just stand up, cross your arms, stand silently," Thorp advised the crowd at the close of another keynote address. "We have to find ways to participate in the liturgy. The liturgy belongs to us."

The same spirit of creeping radicalism rose in the VOTF people at my lunch table on the second day of the conference. Next to me was Mary Ellen Siudut, who seemed like an affable, middle-aged, Catholic mom. She and another woman at the table were members of the Boston affiliate of VOTF. At their last meeting, the focus was to be how to get involved in parish councils, VOTF's recommended "structural change" activity. But these women and others hijacked the meeting, convincing more than half the people there to talk instead about how to respond to Archbishop O'Malley's refusal to wash women's feet at the Holy Thursday Mass.

They are still furious about it. Over lunch, they begin to plan a public protest. They will go to church but stay outside. They will keep their children outside with them. They will gag themselves by putting white masking tape across their mouths. They will carry posters that say "We have no voice in this Church." Other posters will have quotes from women theologians. And they picked the date for their action: Mother's Day.

I felt like I was literally watching the center shift. But in the end, to my great disappointment, that planned action proved too radical for the Boston VOTF affiliate. What a sad commentary, I thought. I had just heard that twenty years earlier, such a rejection of women at a Holy Thursday service would have resulted in a "revolution," with Catholic women throwing themselves barefoot on the church stairs, this according to theologian Susan Muto (an author of the U.S. bishops' pastoral on women), who made that comment in another workshop.

But national VOTF did respond to O'Malley's insult, thanks to Peggie Thorp. She proposed an open letter on the website that would denounce O'Malley for his divisiveness and decidedly unpastoral behavior in excluding women from the sacred Holy Thursday ritual. The reception among VOTF representatives was not unanimous; Thorp had to argue for her position. One person agreed to the letter only if it also defended another group of laity being dissed by the Cardinal—gays—by denouncing O'Malley's inflammatory rhetoric regarding gays who marry. That was fine with Thorp. But even then, another VOTF person refused to approve the letter. The majority ruled, and the letter was posted.

It was a small victory but Thorp saw it as a turning point for VOTF. It officially articulated a concern regarding women in the Church. VOTF is trying to change the structure of the Church by getting lay voices to the table, she explains, and those voices include women. "If the Church is going to alienate or marginalize 50 percent of those lay voices, how can we call ourselves a voice for the laity? . . . At some point, we have to stop apologizing for being fully representative of the laity. . . . We need no longer make excuses for using the word 'women' or speaking on behalf of women in the Church."

There was no flak for that press release, which may confirm Thorp's impression—and mine—that the center is slipping left. While acknowledging that most people in the Church are moderate, "very traditional and abiding," she suspects the meaning of the word "moderate" may have changed in the last two years. "A moderate in the Church accepted the male priesthood, for example," she explains. "Today, I think that same population is saying, 'You know, I wonder if that would have happened if women had been ordained. I wonder if this would have happened if maybe there were more laity involved in decision-making.'"

Brooklyn-Queens VOTF leader Melissa Gradel considers herself a liberal on issues like birth control, women's ordination, optional celibacy, and divorce and remarriage but attests to the same shift. "I think it's wrong to label those as progressive opinions," she argues, "as lefty opinions . . . out of the mainstream because I don't think that they are." Indeed, surveys regularly show that the majority of Catholics favor all of those issues. Even self-described conservative Catholics like cofounder and former coordinator of the West Side Manhattan chapter of VOTF Maria Coffey—who, for example, is deeply opposed to abortion—support optional celibacy, women's ordination, and full inclusion of homosexuals in the Church.

The majority of Catholics do not, however, join reform groups or reform activities. A U.S.-based effort in 1997 failed to duplicate the success of the petition drive of International Movement We Are Church, gathering less than fifty thousand signatures. Even after the scandals broke, the combined membership of the largest U.S.-based reform groups—Call to Action and Voice of the Faithful—totaled less than sixty thousand, out of more than sixty million American Catholics.

On the positive side, because the organizations represent widespread Catholic beliefs, they are speaking for a much larger constituency than their numbers show. On the negative side, it may be that ordinary Catholics simply are unwilling or unable to take on the Church. Thorp is surprised that even a centrist organization like VOTF, which does have 210 affiliates in thirty-eight countries and all fifty states, still can't seem to move much past thirty thousand members. But she understands why Call to Action plateaued. "They were too bold, too brave, ahead of their time. They were telling the truth before most Catholics could stand to hear it." She describes Catholics as a "tough breed. . . . You don't move them very easily." She thinks that is "by design," that it grows out of having been "educated and trained in acquiescence."

This is something that Brooklyn VOTF member Melissa Gradel has thought a lot about. She takes VOTF's goal of structural change very personally. The Church needs to be changed, yes, she says, but so must Catholics, who are used to sitting passively by. She marveled at her own fear of speaking out about discrimination against women in the Church when she met with Brooklyn diocesan priests to argue for lifting the ban on VOTF's meeting on Church property. By contrast, she told me, she would have felt entirely justified and empowered to speak if faced with the same blanket prohibition against women's participation in any other institution. "The structural change that has to take place in the Church has to happen here," she says, laying her hand on her heart, "inside us."

The old-guard reformers had the advantage of beginning when the doors to Church renewal were opened to them. Laity came in, with great energy and exuberance, and made a lot of demands. The Vatican and the bishops in the intervening years pushed them out. The old reformers learned to make do; they learned to think of themselves alone as the Church, treating the hierarchy as a distant nuisance that was best ignored. The sex abuse crisis showed what the result of that distancing has been—a more deeply entrenched power structure than ever before. The new reformers worked immediately to change that. They engaged the hierarchy directly, without benefit of an invitation, defending their action by a brilliant appeal to Church doctrine and canon law. They were not so much asking to be let in as demanding their rightful place at the table.

No doubt the new kid on the block, VOTF, has divided the reform movement. Two years after its founding, it continued to dissociate itself from the old-guard groups, to the dismay of people like *National Catholic Re-*

porter editor Tom Roberts. In his closing keynote at VOTF's New York regional conference in 2003, he lauded VOTF's work but made a plea. "I hope . . . you don't define yourselves over and against the Women's Ordination Conference, Call to Action, FutureChurch. They are easy targets. But when no one else would, they kept the questions alive. . . . We don't need any more divisions."

So contentious do women's issues remain in the Church that even strong Catholic women change-makers like these leaders of VOTF have agreed to put aside these issues in order to create a large moderate movement for change in the Church with clear, less controversial goals. It is not just women's issues, however, or only progressive issues, that VOTF leaders are willing to put aside. The organization's leaders on the right, left, and center have made the same decision, in the interest of finding common ground. All attest to the constant challenge of keeping the Church's warring factions focused on VOTF's three limited goals.

But at the same time, the commitment of VOTF's women leaders to women's equality in the Church runs deep, and VOTF provides them with opportunities to act on it. If this organization does indeed give voice to the laity and brings that voice to the public and to the hierarchy, all of the contentious issues that the clerical hierarchy has tried to squelch will make the agenda. "You're going to hear [us] more and more . . . talking out on issues," Troy told me. "You're going to hear our voice in all of its manifestations." She argues strenuously for open discussion in the Church, seeing it as critical to negotiating change. It provides "an opportunity for conversion of hearts," she says, "both ways."

Joan Chittister took some time to issue her evaluation of VOTF and concluded that they were definitely on the right track. "The truth is that to aspire to give laypeople a 'voice' in the ongoing development and direction of the church stands for the biggest issue of them all," she wrote in the *National Catholic Reporter*. "It stands for declericalization," which she defines as the foundation for the Church's renewal. "If the church is declericalized—if the laity really begins to be included in the theological debates, the canonical processes, the synodal decisions of the Roman Catholic church—every issue on the planet will become grist for its mill."

I find it extremely troubling that the fight for dialogue is so desperate to-day. That the days when the hierarchy listened to the laity are long gone, and so much ground has been lost. I am also troubled that Catholic reformers must proceed so gingerly to speak of women who are so vital to the life of the Church. But I believe that the rise of women's voices in VOTF on their own behalf is yet another example of the stubborn reemergence of Catholic women's vision of, and willingness to fight for, a truly egalitarian institution.

For now, one thing is perfectly clear: If fundamental, systemic change is to come to the American Catholic Church, the hierarchy will not be leading the way. That makes the work of America's reform movement, in all of its manifestations, more important than ever. And that means, in large measure, the work of women. Heading the list are the women who came to the Church fathers demanding justice, bearing broken bodies and broken hearts. They include the victims of predatory priests and miscreant hierarchs—as well as the fiery advocates who rose to speak and act on their behalf.

4

SEX, PRIESTS, AND GIRLHOODS LOST

I discovered Rita Milla in 1991. The few facts I knew about what had happened to her at the hands of Catholic priests appalled me. But on a deeper level, I was as appalled by the hypocrisy as I was by the facts—that this Church that so humiliated and demonized Catholic girls and women for their sexuality would not only tolerate but willingly hide the most egregious sexual crimes of its own clergy.

Like many Catholic girls, my passion for change in the Church came out of the pain of my sexual humiliation. I can still hear my nervous adolescent self struggling to confess my sexual "sins"—a French kiss, petting above, petting below—my feeble attempts to make an intimate connection to another human being. It didn't matter that I had just one boyfriend. Father Alfred denounced me as a "slut" and a "tramp" anyway. At the same time, we were taught to revere our priest judges, who were closer to God than we could ever hope to be. Divorced from physicality, they were undefiled by female flesh—on the altar or in their beds. We, on the other hand, had to live in that flesh. The price we paid in diminished self-esteem lasted a lifetime for many of us.

Father Alfred was not alone in his denunciation. Many priests brought to those confessionals a belief in women's immoral nature, using that holy venue to chastise girls and women for their sexual transgressions. Preeminent theologian Diana Hayes—a creator of "womanist" theology, a variation of liberation theology focused on African-American women—observed that firsthand. Hayes studied for a doctorate in sacred theology at prestigious, pontifical Catholic University. In the 1980s, she was the only woman in her class on the sacrament of reconciliation or confession, and the priest professor forbade her to speak. As a teaching method, he role-played one penitent after another, while each male seminarian role-played a priest. All of the penitents the priest role-played were women, and all of their sins were sexual. "I'd come out of that class just feeling like I'd been bathed in dirt," Hayes told me.

In time, what rose in me was a defense of my body, my sexuality, women's sexuality, my right to make moral choices, to listen to my conscience. Up rose a voice of self-respect that came from God, not from my confessors. Things turned around for me—perhaps fittingly—after my last private confession, three decades ago. I had been feeling guilty over my estrangement from the Church. I longed for the Mass and the Eucharist. But first, I would have to go to confession. So I headed for the priest's rectory at Saint Ann's early one morning on a visit to Scranton, hoping to be healed. I was, but not in the way I had expected.

The priest's alcohol-laden breath wafted through the confessional screen. He asked how old I was—I was twenty-four; if I was single or married—I was single. Then he invited me into the parlor to complete my confession, where he sat too close, draped his arm around my shoulders, slobbered, and pleaded for my address in Manhattan. I learned that day that the Church is made of men, and those men are not gods.

One Woman's Story

I suggested a story about Rita Milla to *Ms.*, where I was a contributing editor. *Ms.*'s editor in chief at the time, Robin Morgan, feared a charge of Catholic-bashing if I looked only at Milla's story. So she suggested I investigate clergy sex abuse across denominations, which I did. I found sexual exploitation among Buddhists, where Zen teachers made sex a requirement for their female students' spiritual instruction. I found vulnerable women at times of deep crisis in their lives who were the objects of sexually rapacious Jewish and Protestant clergy counselors. But I found nothing that compared to what I learned about Rita Milla.

A Latina who grew up in Los Angeles, Milla was a devout Catholic. At Saint Philomena Church, she went to confession every Saturday and Communion every Sunday. She sang in the choir. She cleaned the altar. That is where she met Father Santiago Tamayo.

Twenty-seven years her senior, Tamayo paid a lot of attention to Milla. That made her feel special, she told me, as if "God turned around and looked at you." In 1977, when Milla was sixteen, Father Tamayo began to make sexual advances toward her, reaching through a broken confessional screen to touch her breast. When she turned eighteen, those advances included intercourse. Milla said that for the next several years, Tamayo facilitated sexual liaisons for her with six other priests. One night, Tamayo and Father Angel Cruces took her to an empty house. On another night, four priests, including Cruces and Tamayo, took Milla to a hotel room that they rented by the hour. On still another occasion, Milla had sex with two other priests in Tamayo's rectory room. Shortly thereafter, she found herself pregnant.

Tamayo suggested an abortion but Milla refused. Though Tamayo was not the father, he sent Milla to his homeland, the Philippines, where she was to give birth and leave the baby behind. He told her parents she was leaving to study. Things did not go as Tamayo had planned. Milla developed pregnancy complications and nearly died. She finally told her family; her mother and sister traveled to the Philippines to take her home.

When Milla, with her newborn daughter, returned to California, she and her parents went to see the late Los Angeles Bishop John Ward. She said he offered no help at all. When I called him years later, he spoke of her with the utmost disdain. "She was referred to the proper authorities when she came to my office," he said. "That was the end of it. I had nothing more to do with her." Disillusioned, on welfare, and living with her daughter at her parents' home, Milla decided to sue. In 1984, she turned to fiery feminist attorney Gloria Allred, who promptly filed a multimillion-dollar lawsuit against the seven priests and the Los Angeles archdiocese, making Milla one of the first Catholic women to do so. As soon as the lawsuit was filed, Allred reported, all of the priests mysteriously disappeared.

The case was thrown out of court in part because it exceeded the statute of limitations. That could have been the end of the story, had it not been for Tamayo's conscience. In 1991, he returned to Los Angeles to publicly apologize to Milla and her family. In his apology, Tamayo also reported: "When the scandal became public, I followed the orders of my religious superiors. I immediately left my parish as instructed and made myself unavailable to the press and process servers." To the chagrin of the Los Angeles archdiocese, he released supporting documentation.

The archdiocese, in writing, had instructed Tamayo not to return to Los Angeles to face the suits against him; advised him to become a parish priest in the Philippines, in spite of the unresolved charges against him; and promised him money in the form of a donation, with the stipulation that he not reveal its source unless forced to under oath. The most powerful men in the archdiocese were among the signatories or recipients of these letters, including Cardinal Roger Mahony, now head of the Archdiocese of Los Angeles, the nation's largest.

When I interviewed Tamayo, he had been banished from the priesthood not for sexually abusing a minor but because he had gotten married. He told me that one of the other priests involved in the sexual abuse had been comfortably ensconced for years in a parish in Brooklyn—then my own backyard. I went one day to watch Father Angel Cruces serve Mass. I found a vigorous old man with a booming voice, resplendent in a brilliant green vestment. The sight of him made my stomach turn. After Mass, I went up to the altar and into the sacristy to talk. I said I was a reporter doing a story on Rita Milla. His face flushed. He began to hyperventilate. "The case is in the past. Talk to my lawyer," he said, rising. As I left, he slammed the door behind me, never giving me his attorney's name.

The Brooklyn diocese, then headed by Bishop Thomas V. Daily, maintained that when it accepted Cruces in 1985 as a guest priest from the Philippines, it knew nothing about the accusations against him. In fact, said spokesman Frank DeRosa, the diocese did not learn of the charges until the middle of September 1991, six years after his arrival and within days of my inquiry. DeRosa insisted that new paperwork from the Philippines, not my call, resurrected the charges—a confluence of events I questioned, which even DeRosa declared to be "very coincidental, absolutely coincidental . . . it's unbelievable, it really is."

Before I knew it, Cruces was gone. Though that was entirely in keeping with what I knew to be the hierarchy's strategy, I was stunned. Ten years later, in 2002, when the clergy sex abuse scandals hit the newspapers, Cruces was discovered back in New York, serving at Holy Cross Church in Manhattan. After more bad publicity, he was again reassigned, back to the Philippines. In 2003, Bishop Daily resigned, amid angry charges that he had failed to protect children from known priest pedophiles, both in the diocese of Brooklyn and, decades before, in Boston.

Girl Victims Invisible

When the clergy sex abuse crisis burst onto the scene in 2002, the abuse of girls by Catholic priests remained invisible or, at most, a footnote. Yet the crisis has never been only about the abuse of male minors.

Father John Geoghan was the serial pedophile at the center of the storm in the Boston archdiocese when the sex abuse reports first broke. Survivor and activist Susan Gallagher made the point early on that, while most people had no idea, females were among Geoghan's victims. In fact, twenty-three out of his 148 victims were girls—sixteen to whom he exposed himself and seven of whom he molested. The attorney who represented those victims, Mitchell Garabedian, estimates that nearly all of Geoghan's victims were poor, and 90 percent of them were being raised by women alone. Indeed, Gallagher herself—one of seven children raised by a struggling, working mother who was abandoned by her husband—was preyed upon by an equal-opportunity pedophile. The Reverend Frank Nugent sexually molested her and two of her brothers—an agony that led to the death of one of her brothers in what Gallagher calls a suicidal car crash. So disturbed was Gallagher by the *Boston Globe*'s omission of stories concerning female victims until the end of 2002 that she launched a website to address that omission.

Of the ninety-nine victims of notorious pedophile James R. Porter, former Fall River, Massachusetts, priest, twenty percent were female. Some pedophiles have abused scores of females: Father Robert E. Kelley admitted to sexually abusing up to one hundred girls, and Father Carl Warnet preyed on girls as young as eight, fifty of whom were identified by name, according to attorney William Crosby, who worked on that case. "When you look at the serial pattern offenders, you'll see a preponderance of male victims," observes attorney Jeff Anderson, who has represented more than five hundred survivors of priest sex abuse. "But if you look closer, you'll always see a female victim because it's a matter of power and the abuse of it."

Experts on clergy sex abuse have estimated that some 30 percent of the victims of priest child sex abuse are girls. However, the survey commissioned by the United States Conference of Catholic Bishops and conducted by the John Jay College of Criminal Justice on the incidence of child sex abuse by priests and ordained deacons found a lower incidence. It revealed that more than ten thousand children had been sexually abused by nearly 4,400 clerics, or 4 percent of the nearly 110,000 Catholic priests reported to have been serving in the United States from 1950 to 2002. Nineteen percent of those victims were girls. However, the gender of the victim was unknown for nearly 10 percent of the priest abusers. And as for the priests who did the abusing, a total of 26 percent either abused girls alone, or boys and girls.

More than three-quarters of the victims did not meet the clinical definition of pedophilia, which refers to a prepubescent child under the age of eleven. The survey found the largest number of girls were abused between the ages of eleven and fourteen, as were the largest number of boys. From ages eleven to seventeen, boys outnumbered female victims by nearly six to one. But more girls than boys from the ages of one to seven were abused— 40 percent more; while this was a small category, it was also the only category where the number of girls exceeded the number of boys.

Many believe the number of minor victims reported by the John Jay survey represents a significant underestimate. First, the bishops themselves provided the information. As activist survivor Claudia Vercellotti put it: "To rely on the diocese self-reporting is about as practical and reliable as relying on Ken Lay and the Enron accountants to unearth the severity of the Enron scandal." Another factor is the severe underreporting of crimes of sexual abuse, to which the John Jay survey itself attested. Psychologist Gary Schoener—who has counseled victims and priest abusers, provided expert testimony, and assisted in the development of church sex abuse policies—points out that about 3 percent to 5 percent of victims report their abuse to anybody in authority. Based on that underreporting, some experts have put forth their own estimates of minors abused by Catholic priests, ranging from twenty thousand to one hundred thousand victims in the United States alone. Finally, the vast majority of cases reported by the bishops' staffs to the researchers occurred from the 1960s through the mid-1980s. Because it takes years for victims to come forward, even the researchers cautioned that additional allegations from the survey period could surface in the future—and have.

Little Eyes

As concerns the much lower incidence of female than male victims, plaintiffs' attorneys report that men bring the vast majority of cases for child sexual abuse by Catholic priests; an estimated 95 percent of those who filed suit in Boston were men. However, that may have less to do with the actual incidence of abuse than with the reluctance of women to come forward, and that has to do with how women victims are treated and seen.

Psychologist Gary Schoener and many other experts have observed a dramatic difference between the treatment of women and girls who are sexually abused and the treatment of men and boys. "In a deposition, if you've got any [female] over the age of twelve, you can bet she's going to be called seductive. It'll come up dozens of times," he explains. "By contrast . . . I've never heard a boy accused of being seductive, ever. . . . The issue of holding the woman accountable or blaming her, that's there. It's lurking just below the surface, and it presents a tremendous obstacle to overcome." In that sense, the sexual abuse of teenage girls by priests has been viewed "much the same way as rape has been viewed in the past," said an activist survivor, referring to female culpability.

Indeed, while the Church has barely a leg to stand on when the abuse involves a small child, male or female, with girls approaching puberty and be-yond, the rationalizations for the priest's behavior begin to fly. Nancy Sloan was just this side of puberty when she met Father Oliver O'Grady at a Catholic camp near Nevada City, California. He fussed over her. Befriended her parents. Came to dinner. So when he invited her to spend a weekend with him at his home forty-five minutes away, her parents were delighted. "By today's standards, people would say, 'That's weird,'" says Sloan. "Not at the time. He was a priest. He was to be trusted. My mother would have never thought otherwise."

The consequences were dire. For four days, O'Grady molested Sloan. "The last memory I had with him is severe pain while he was trying to pen-etrate an eleven-year-old," she says. When O'Grady suggested another sleep-over, Sloan told her parents what had happened. Devastated, they contacted O'Grady's parish. His pastor, Cornelius DeGroot, confronted O'Grady, who admitted to the molestation. DeGroot advised O'Grady to write a letter of apology to Sloan's parents, which he did. In his letter, O'Grady admitted to "touching Nancy in areas I shouldn't have" and that he "went a little too far." It was an amazing rationalization of his crime, raising the question of just how far is *not* too far for a grown man to go with an eleven-year-old girl.

In 1994, Sloan, accompanied by Debra Warwick-Sabino, a victims' advo-cate and pastoral counselor, met with authorities of the Stockton diocese, including Monsignor James Cain. Sloan had learned of another allegation against O'Grady and wanted to know why the diocese had taken no action since she made her original complaint more than fifteen years before. She also wanted an apology. According to Sloan and Warwick's handwritten notes of the meeting, "Monsignor Cain . . . said that 'he and Nancy spoke a different language.' He didn't apologize. He said that because she was fe-male, he didn't realize how serious it was—that O'Grady may just have been curious and if she were a male child, he would have taken stronger action against O'Grady."

Rita Milla was held responsible by some for having been turned effec-tively into a prostitute by Father Santiago Tamayo, who had begun to groom her at the age of sixteen. When she later confessed to a priest about her involvement with Tamayo, hoping for absolution and help, instead, she reports, "The priest gave me this lecture on how some women enjoy making priests sin. I thought, I must be one of those women."

Children at or around puberty and teenagers are at greatest risk of being locked into these sexually abusive relationships, with no way out. "The developmental stage at which a child is abused . . . effects [*sic*] their ability to reject the perpetrator and to disclose sexual abuse," read the report of the Suffolk County Supreme Court Special Grand Jury, which investigated sexual abuse of minors by priests in the diocese of Rockville Centre, Long Island. Based on the work of its psychological consultant, Dr. Eileen Treacy, it said: "The children who are least likely to disclose, and the most difficult to treat, are grammar-school age. This is because they are abused as they are approaching puberty. Adolescents are not far behind simply because they understand the consequences of disclosure on a more sophisticated level."

The implicit belief that teenage girls are in a sense fair game for Catholic priests was unwittingly expressed by the powerful Chicago Cardinal Francis George. In April of 2002, under the hot lights of public scrutiny, he argued that not all priests guilty of sexual misconduct should be defrocked or laicized, which ends their ministry while also returning them to the lay state. In a banner story in the *New York Times* on the pope's summoning of America's cardinals to Rome to talk about the sex abuse crisis, George was quoted as saying, "There is a difference between a moral monster like [Boston's serial pedophile] Geoghan and an individual who, perhaps under the influence of alcohol," behaves inappropriately "with a sixteen- or seventeen-year-old young woman who returns his affections."

It is a shocking defense, deeply revelatory of the Church's view of women and female sexuality. The abuse of a female minor after puberty is not pedophilia, but it is deeply damaging. In Cardinal George's example, the young woman is entirely invisible. There is absolutely no awareness of the potential damage to her—psychologically, spiritually, physically, or materially. The Church condemns all sexual activity outside of marriage and all forms of artificial birth control and abortion. The priest in George's example would be transgressing on all of those grounds, potentially leaving the young woman with an unwanted pregnancy and a child to support, most likely condemned to a life of struggle and economic destitution, for which the Church and its patriarchs have repeatedly refused to take any, or only the most begrudging, responsibility.

When Cardinal George describes a girl who returns the priest's "affections," he does not see that girl as a minor but as a consenting adult, wholly capable of a clear-eyed assessment of the risks and benefits of involvement with a priest and the ability to decide to go ahead anyway. While there may be a teenager here or there who could do that, the laws forbidding sex with minors have been written to protect the vast majority who cannot.

So extreme is the hierarchy's refusal to see older teenage girls as victims that even in the face of the massive public outcry over clergy abuse of minors, Bridgeport, Connecticut, Bishop William Lori defended his decision in 2003 to leave in ministry Monsignor Martin Ryan. As a parish priest back in the 1970s, Ryan had been accused of attempting to force himself on a teenage minor, the parish organist, by inviting her into the rectory, grabbing her by the wrist, cornering her, pressing his body up against hers, forcibly French-kissing her, and groping her through her clothes before she was able to break free and run away—charges he has consistently denied. Bishop Lori made that decision after agreeing to include that victim in the diocese's $21 million settlement with thirty-nine other victims, and after the diocese's own investigation turned up two other girls—a high-school senior and a college sophomore—who said that they had similar experiences with Ryan. "He's not a predator," Lori told the press. "These allegations are old. He's not at all a threat to young people." In announcing the diocese's decision, spokesman Joseph McAleer allowed that Ryan "may have had celibacy issues." The diocese sent Ryan for psychological counseling after the complaint was made in 2002, then returned him to ministry as pastor of Saint Edward the Confessor Parish in Connecticut. "He is entitled like anyone else to his good name," wrote McAleer.

Former Benedictine monk and psychotherapist Richard Sipe, who has studied priests and sexuality for twenty-five years, sheds some light on that line of thinking. For centuries, he says, the Church has worried not about the scandals of priests having sex with children but about priests having sex with boys. Supported by laws that permitted the marriage of very young girls, the Church has failed to see the difference between a priest having sex with a woman and a priest having sex with a girl, viewing both as variations of "normal" sex between a male and a female. I would add that the Church at times also fails to see the difference between forced sex and consensual sex. The way the Bridgeport diocese described the alleged incident in its press release is telling. "She said that Monsignor Ryan tried to kiss her once, touched her, and was rebuffed," it read, making the priest's actions sound almost benign.

That refusal to see priest sexual abuse of teenage girls as a crime may well have influenced the incidence of the abuse of females reported to Church authorities. The only category in the John Jay survey where the abuse of girls exceeded boys was with girls from the ages of one to seven. If Church officials are still so obviously unwilling to see older teenage girls as victims, then it seems plausible that diocesan authorities sent that message to girl victims and their families in decades past, making it extremely difficult for families to report the abuse of older female teens. In addition, some Catholic parents, devoted to the Church, might have been quicker to minimize an incident involving an older daughter than an older son, more willing to give the priest the benefit of the doubt, or, in extreme cases, willing to blame their daughters for what occurred.

Finally, in his great sympathy for an inebriated priest having sex with a sixteen- or seventeen-year-old girl, Cardinal George apparently overlooked the prevalence of alcohol and substance abuse problems, which the John Jay survey put at nearly one-fifth of all sexually abusive clergy.

Observing the extent to which women's complaints of sexual abuse have been minimized, denied, and ignored by Church officials and review boards, attorney Roderick MacLeish, Jr., who represented hundreds of victims of sexual abuse in the Boston archdiocese, told the *Boston Globe*: "It's almost a free pass when it comes to women and young girls." That follows, really, in a Church where male clerics remain the godlike rulers and women—in their private lives and the public life of the Church—the most fiercely ruled. It also follows in a Church that sees women as second-class citizens, to be patronized and managed without voice, and that tends to see "womenandchildren" as a single entity, dismissable in a single swoop.

Women Take Action

There is some irony, then, in the fact that Catholic women have been leaders in calling the Church to accountability on the subject of the clergy sexual abuse of children. Without Suffolk Superior Court Judge Constance Sweeney, the Boston crisis of 2002 might never have broken. Believing that the public had a right to know, and acting on a request from the *Boston Globe*, Sweeney broke with tradition and ordered the release of thousands of documents of alleged sex abusers, which revealed the Church's secrecy, duplicity, and collusion in failing to protect children and keeping child molesters in the ministry, at unspeakable cost. Chosen by the *Boston Globe* as one of the people who made a difference in 2002 in the sex abuse crisis in the Church, they reported that Sweeney was "a practicing Catholic, the product of sixteen years of parochial schools," and it may be that because of that background, "her ruling, unleashing the first tidal wave of disclosures that would eventually bring Law down as Boston's archbishop, seemed earth-shattering."

Some of the women change-makers were given recognition and authority by the bishops themselves. Mary Gail Frawley-O'Dea stood up before the bishops at their historic meeting in Dallas in 2002 where they invited experts, prominent Catholics, and survivors to talk to them about the crisis. A psychologist who specializes in adult recovery from child sex abuse, Frawley-O'Dea spoke about the horrors experienced by sexually abused children, taking the bishops "on a tour of the corridors of the psyche twisted by sexual transgression." She told the bishops in no uncertain terms that what they were facing was "incest" and asked that each of them "reflect on your role in the devastation" of abuse perpetrated by their fellow priests.

No puppet of the bishops, six months later she led eighty-three mental health professionals in challenging a decision by Cardinal Law's temporary replacement, Bishop Richard G. Lennon, to allow church lawyers to interrogate counselors treating victims of clergy sexual abuse who were suing the diocese. "I have a lot of empathy for the bishops who are trying to make things right, and I don't consider the church my enemy," she told the *Boston Globe*. "But I think that this is very despicable and deceitful. To say, 'The church loves you' and 'We want to help you,' and then to invade your treatment is really wrong. It may be legally okay, but it's wrong." In response to the stinging grand jury report of Attorney General Thomas F. Reilly on the Boston archdiocese's horrendous history of child abuse, cover-up, and his call for an independent auditor of the Church's handling of child sex abuse cases, Frawley-O'Dea joined a volunteer group of psychologists, social workers, and lawyers specializing in sexual abuse. They petitioned Law's successor, Archbishop Sean O'Malley, to become that independent auditor. Though the archbishop refused to accept them in that role, the group formed its own independent entity, the Victims' Rights Committee, to review the archdiocese's progress in following its own new policies in the handling of sex abuse cases. Frawley-O'Dea served briefly as its first chair.

Kathleen McChesney, executive director of the U.S. bishops' Office of Child and Youth Protection, managed the first round of audits to monitor compliance of each diocese with the bishops' new charter on the handling of child sex abuse cases. Investigators with the Gavin Group of Boston visited 191 of the nation's 194 dioceses to collect information, while McChesney traveled around the country gathering her own feedback. She met with Church personnel, victims, and any Church reform group that asked—from Call to Action to Voice of the Faithful, to the consternation of some conservative Church officials, like Archbishop John J. Myers of Newark, New Jersey, who has banned VOTF from meeting on Church property.

There was much to criticize in the "Report on the Implementation of the Charter for the Protection of Children and Young People," which was released in December 2003. Most important, each audit reported the diocese's compliance with the bishops' new policies for only one year, from the time of the adoption of the charter in 2002. It thereby cut off at the knees any hope of evaluating how each diocese handled cases in decades past. This meant, for example, that a diocese that had been notorious in moving child sex abusers around would have been deemed one hundred percent "in compliance" with that article of the charter if they had not done so from 2002 to 2003. This inspired a skeptical reading of the hierarchy's self-congratulatory declaration that 90 percent of the dioceses audited were found to be "compliant with all provisions of the charter."

But McChesney built strong observations and recommendations into her report, which some activists deemed its most important contribution. They ranged from chastising dioceses that released perpetrators never brought to justice back in the community, without accountability or follow-up; to calling for Church officials to report *all* cases of child sexual abuse to legal authorities, without regard to statutes of limitations; to including in the policies a requirement that parish priests, as well as bishops, reach out to victims of priest sexual abuse; to insisting that the bishops obtain more input from victims abused before the charter was adopted—the most radical activist survivors who were effectively closed out of the first audit process.

Illinois Court of Appeals Justice Anne Burke was one of the thirteen members originally appointed by the bishops to the National Lay Review Board. When the first chair, former Oklahoma Governor Frank Keating, was fired for publicly venting his frustration over the refusal of powerful cardinals to cooperate with civil authorities, likening their tactics to "*La Cosa Nostra*," Burke became interim chair. A devoted and traditional Catholic, as well as an outspoken defender of the Church hierarchy and Church teachings, she turned out to be no pushover.

Under her stewardship, the board issued the findings of the John Jay survey and the board's own anecdotal report, based on interviews, on the causes and context for the crisis. The board accused the bishops of poor screening of seminary students in decades past, and poor seminary training, especially regarding sex and celibacy. It criticized the bishops for favoring abusive priests over the victims to save face; relying on secrecy and legal adversarial advice; duplicity in imposing a zero-tolerance policy on the priest abusers but not on themselves; and a "troubling" lack of any expression of outrage over the "abhorrent acts of some priests." The board boldly linked the "don't rock the boat" attitude among the current cadre of bishops—which led to their failure to call their own on child sex abuse—to the types of bishops who have been promoted during this papacy. They found that "outspoken priests rarely were selected to be bishops, and the outspoken bishops rarely were selected as archbishops and cardinals." To diversify the leadership pool, the Lay Review Board argued strenuously for opening up "the type of priests who are chosen as bishops by the Holy See" and for involving the laity in the selection of those bishops. And the board called for greater lay consultation overall. "The exercise of authority without accountability is not servant leadership," they wrote. "It is tyranny."

In 2004, the board found out that behind their backs, some of the most powerful bishops were sending letters to Bishop Wilton D. Gregory, president of the U.S. Bishops' Conference, trying to effectively renege on their commitments and criticizing the board as well as the Office of Child and Youth Protection for behaving with too much independence and autonomy. In response, Burke sent a blistering letter on behalf of the Lay Board to Bishop Gregory. She called the bishops on trying to delay the second compliance audit; on putting off approving funds for a scientific study of causes and context; and on failing miserably at "fraternal correction"—that is, at taking action against offending bishops. She accused them of being "disingenuous," of going back to "business as usual," of delaying "healing" and reopening "wounds," of manipulating the board, and of believing that because of the board's work, they thought they had "dodged the bullet" of national scrutiny. The letter demanded an immediate vote on the board's recommendations and insisted on an answer within just a few days.

Chagrined and threatened, some bishops sent angry replies to Gregory. They insisted that the board should have nothing to say on the matter of "fraternal correction among bishops," and that the board was an "advisory body at the service of the bishops," without "supervisory authority." The whole matter became public, illustrating the extent of the power struggle under way in the Church between the laity and the hierarchy, and between powerful laywomen and the bishops. In the public eye, the bishops acquiesced; the second compliance audit commenced, as did plans to request proposals for the conduct of a large scientific study of the causes and context for the sex abuse crisis.

Ordinary Catholic women were also shaken out of their slumber by the horror of the revelations rolling across the front pages of the nation's newspapers. Anne Barrett Doyle was among them. Mother of two and a cradle Catholic, Doyle "literally woke up" in January 2002, when she read the *Boston*

Globe's breaking stories of clergy sex abuse and diocesan cover-up. The same spirit that had moved her as a tenth grader in a packed Roman Catholic Church to stand up and protest her priest's refusal to baptize the baby of pro-choice parents inspired her again. Instead of going to their usual parish, she and her family drove to the Cathedral of the Holy Cross, where Cardinal Bernard Law would be serving Sunday Mass. Carrying her scrawled "It's My Church" sign, Doyle joined a picket line of seven. "I was so overcome with the sinfulness of the Church and also of myself, as a layperson who had enjoyed being part of this little club and had not fought against [my] subservient role," says Doyle.

Her wake-up call resulted in the birth of one of Boston's most spirited Church reform organizations—the first ever to join lay Catholics in public support of survivors of child sex abuse by Catholic priests—the Coalition of Catholics and Survivors. It was the coalition, in communion with raging survivors, that organized the earliest marches and demonstrations in response to the scandals. On Good Friday, in 2002, hundreds of the faithful linked arms and literally encircled the cathedral, a symbolic representation of Law's predicament from which, in the end, there was no escape.

At that event, survivors were invited to the podium. One of them was Barbara Blaine. In the long years since she had founded the Survivors Network of those Abused by Priests (SNAP), she had never before seen Catholic laypeople or a Catholic organization stand publicly with survivors. She told them how moving that was for her. Indeed, that was a time when survivors were truly lone voices, with little support from anyone, not the hierarchy or ordinary Catholics in the pews.

Blaine's Battle

Barbara Blaine grew up a good Catholic girl. She labored happily around Saint Pius X parish in Toledo, Ohio, with a group of other seventh- and eighth-grade girls who were dubbed by the priests as "the deaconettes." One priest had become particularly close to her family—Chet Warren, an order priest with the Oblates of Saint Francis De Sales. One night, he invited her to stay for dinner at the rectory after she had cleaned up from a baptism. She sped home to ask her parents' permission, which, of course, they gave. After dinner, the other three priests left, and Warren took Blaine into the TV room. But pretty soon, she says, "Warren was watching me, not the TV." That night, the sexual molestation began. For the next four years, Warren sexually molested Blaine, the assaults ranging from groping to digital vaginal penetration. He blamed her for the abuse. He said his erections were her fault, that she was "irresistible," and he couldn't help it. The message Blaine got was that "being a woman is bad. It's evil. It's making this happen to him."

In time, Blaine became physically sick, developing severe migraines and even being hospitalized. Her abuse did not end until she was a senior in high school. It was after she had her first face-to-face confession. When she told the priest about Warren, his response was: "Barbara. Jesus loves you. Jesus can forgive anything." "As sick as that sounds today," she says, "it gave me a sense of self-esteem to go back home and to tell Chet Warren he can't do this anymore." From that time until age twenty-nine, Blaine kept the secret of her abuse. She remained a devoted Catholic. She volunteered with the Sisters of Mercy, entered and left the convent, and earned a degree in social work. She worked for the Catholic peace organization Pax Christi, for Chicago's Catholic Worker House, and for a homeless shelter. She also began to study for her master's in divinity part-time at Catholic Theological Union.

Then in 1985, Blaine read about the sex abuse scandals in Louisiana, uncovered by journalist Jason Berry. "I began to experience symptoms of posttraumatic stress disorder. I was angry. I would cry for no reason. I was having flashbacks and nightmares. I was being haunted," she says. Edwina Gately, an independent Catholic minister beloved by the progressive Catholic community, used to come a few times a week to the homeless shelter where Blaine worked to talk to the prostitutes. Founder of a program for prostitutes called Genesis House, Gately was one of the first people Blaine told what she was going through, and "the only one who had any insight," says Blaine. What Gately told Blaine was: "Barbara, that's incest." At Gately's suggestion, Blaine joined a support group for incest survivors.

That same year, Blaine made her first contact with the late bishop of the Toledo archdiocese, James Hoffman, and the former provincial of the

Oblates, to tell them about the abuse. They feigned ignorance, she reports, but in fact, they already knew. By then, the diocese had had complaints from two other girls about the sexual abuse by Warren; the transfers of Warren had already begun. For seven years, Blaine pursued representatives of the diocese and Warren's order for help, traveling back and forth from her home in Chicago to Toledo on her own dime. She asked for money for counseling, which she eventually got, but only after fighting a six-session limit and the diocese's outrageous insistence that Warren attend the sessions, too. She pleaded with diocesan officials to protect other potential victims by monitoring and treating Warren, which they said they would do.

Blaine also went to her family, which was supportive; to the police, who said it was too late to file charges; and finally, to other survivors, forming SNAP in 1989. She held the first meeting at the Holiday Inn in Chicago, which drew a small but enthusiastic group of survivors from around the country. She contacted people like journalist Jason Berry, who linked her to other survivors. She set up a dedicated SNAP phone line at the Chicago Catholic Worker House. She watched the network grow, as survivors volunteered to hold support groups in their hometowns. And she worked with the nascent SNAP chapters to organize actions; from the earliest days, SNAP members picketed and leafleted at parishes and chanceries, outing abusive priests.

In the meantime, Warren had been transferred to a common dumping ground for criminal priests, a hospital chaplaincy office. The last straw came when Blaine's father was hospitalized there, and Blaine found out that Warren was not being monitored at all. Incensed, Blaine met with representatives of the diocese, Warren's order, and the hospital's risk management department to demand action, to no avail. Soon after, she was invited to appear on *Oprah*. She contacted all the same people to tell them the news. This time, they listened. Within days, Warren was removed from the ministry and, later, dismissed from his order. That was in 1992. "That's what it took," says Blaine.

Soon after, the *Toledo Blade* broke Blaine's story. Some three weeks later, on January 3, 1993, Oblate Provincial Reverend James F. Cryan released a statement confirming Blaine's charges. He said that Warren denied them. While making no judgment whatsoever about Warren's character, he publicly pronounced Blaine "clearly a troubled and anguished person."

The hostility toward activist victims of priest sex abuse like Blaine persisted well after the Dallas charter was signed. Late in 2002, when Blaine was about to speak in her hometown on behalf of SNAP, the *Toledo Blade* asked Reverend Thomas Quinn, then director of communications for the Toledo diocese, for a comment. "Where do we place the bombs?" he asked. "And you can quote me on that."

By 2004, SNAP had grown to sixty chapters nationwide and 5,100 members—half of whom are women. The year before, Blaine, then forty-eight, left her job as a child welfare lawyer to devote herself full-time to being SNAP's president, traveling nonstop to meet with survivors and speak at the conferences of sympathetic Church reformers, from Call to Action to Voice of the Faithful. In the spirit of Barbara Blaine, SNAP has become a channel for the anger and hurt as well as the energies and hope of survivors nationwide. Rita Milla is the Spanish-speaking contact for survivors and press in Southern California and a regular at press conferences and public events, some of which her daughter has also attended. Survivor Nancy Sloan is the media contact and an active organizer for the Sacramento SNAP chapter, an outspoken and energetic advocate for victims.

Today, as always, the lion's share of SNAP's work is supporting survivors, who call, e-mail, visit SNAP's website, and join local SNAP support groups, seeking advice, solace, and a place to take a lifetime of secret pain. And SNAP remains relentlessly vocal, bold and unself-conscious in its demands that offending priests and the hierarchy who covered for them be held accountable for their crimes—which they see as essential to the healing of victims and of the Church. In that regard, SNAP points fingers and names names. Its members protest and march and leaflet all over the country at churches and chanceries, bringing the information they uncover to parishioners, the hierarchy, and the media.

Those protest actions used to be lonely, even dangerous, undertakings.

That's not so much the case anymore. "The clearest sign of progress is from the bottom up, not from the top," observes David Clohessy, SNAP's executive director, who spoke for SNAP at the 2004 bishops' meeting in Dallas and whose tearstained face on newspapers across the country became the symbol of the suffering of priest child abuse survivors. "We have been leafleting for more than a decade in front of churches. I had a parishioner spit in my face. I saw a woman thrown down the steps. Now, things happen, but they're pretty minor. Now, people take a flyer, walk away, stop, turn around, walk back, and say: 'I'm glad you're here. Can I give you a hug? I'm sorry for what happened to you.'"

SNAP members are helping to extend or abolish statutes of limitations for child sex abuse cases, which pose the single greatest impediment to a victim's ability to bring their perpetrators to justice. In various states, SNAP members have testified about the time it takes for victims to come to terms with the abuse and find the courage to come forward, which can be as long as twenty-five years. While more than half the states now recognize "delayed discovery of injury"—that is, the statute of limitations on child sexual abuse does not take effect until the victim is old enough to recognize the damage done—many states still require that charges be brought within a certain number of years of the age of maturity, usually starting at eighteen years old.

In the battles to change these laws, victims' advocates are often pitted against the Catholic Church's most powerful lobbyists, who vociferously defend maintaining current law. With SNAP's help—particularly Nancy Sloan's testimony—the most dramatic change came in California in 2003, when the state suspended the statute of limitations in civil cases for one year for survivors of child sexual abuse. Any victim, regardless of how much time had passed since the abuse took place, would be able to bring civil charges against the perpetrator and the institution that harbored him. Within that twelve-month window, an estimated eight hundred cases were filed—five hundred of them in Los Angeles—and the majority were against the Catholic Church.

SNAP keeps an eye on the tactics being used by the Church hierarchy in court, such as continuing to defend against lawsuits by insisting that the victim's parents should have known better, and staffing the diocesan victims' phone line with their own defense attorneys—tactics SNAP discovered were in use in Kansas City and Brooklyn, respectively, in 2004. And SNAP helps victims to get their counseling paid for, fights diocesan-imposed limits on sessions, and advocates for third parties, not the diocese, to handle counseling arrangements.

And SNAP is ever on the lookout for priests remaining in the ministry who should be out. In 2003, a year after the Dallas charter was signed, it was SNAP who publicized the case of a priest in Santa Ana, California, who was accused of having child porn, but the diocese claimed that behavior was not covered by the bishops' charter. SNAP also found and publicized the case of another priest who was actually living in Chicago's Cardinal Francis George's residence one week a month to help out with liturgy, who it turned out had pled guilty in 2001 to child molestation charges. The diocese claimed the prior charge was "irrelevant" because it occurred "before he was ordained." And it was SNAP that called national attention to the case of Monsignor Martin Ryan, left in the ministry by Bridgeport Bishop Martin Lori. SNAP leaders urge Catholics to remain vigilant, and to evaluate the behavior of the Church hierarchy not by the Church's standards but by using their own common sense.

In 2004, SNAP lent their support to another class of survivors: men and women abused by nuns. Led by SNAP leader Landa Mauriello-Vernon, who says she was abused by a nun in high school and the novitiate, a small group of victims made an appeal for time on the agenda of a joint meeting of representatives of the 100,000 U.S. Catholic priests, sisters, and brothers who are members of the Conference of Major Superiors of Men and the Leadership Conference of Women Religious. In a response reminiscent of the days before the bishops signed their Dallas charter, the leadership conference refused SNAP's request, despite the fact that the subject of the conference was violence, from domestic abuse to war. In a press release, the leadership conference said the meeting was not "an environment conducive to listening and dialogue." It was a troubling response from a historically progressive women's organization.

New Directions

Since the sex abuse scandal broke in 2002, Catholic women have responded in some creative new ways. Anne Barrett Doyle, cofounder of the Coalition of Catholics and Survivors and an advisor to Voice of the Faithful, is also codirector of *BishopAccountability.org*. Created by advocate, editor, and document manager Terence McKiernan, it is the largest source on the Internet of comprehensive historical information about the Catholic sexual abuse crisis and the bishops' responsibility for it. Included are links to newspaper investigations, attorney general and grand jury testimony, and reams of court documents, making it an unparalleled permanent archive.

The Linkup is an ecumenical organization for clergy sex abuse survivors, founded by a woman, Jeanne Miller, in Chicago. It has three thousand members nationwide, half of them women. Its president, Susan Archibald, spearheaded the creation of the first national treatment facility for clergy abuse survivors. Opened in June 2004, The Farm is a wellness center located on 1,300 acres outside of Louisville, Kentucky. It provides a host of alternative therapies (from yoga to organic gardening to bread making to art therapy), as well as mental health treatment (for depression, anger management, self-esteem building, etc.), lectures, workshops, and retreats. Beyond The Farm's obvious meaning to survivors, Archibald sees it as a bridge between survivors and the Church hierarchy, who she believes are more deeply estranged than ever before. For months, Archibald worked to "develop diplomatic relationships with the Church," meeting with bishops and heads of religious orders to encourage them to participate in and support the center, including financially. By the fall of 2004, thirty bishops and cardinals, as well as forty members of male religious communities, had donated to The Farm.

And a group of therapists, spiritual directors, and survivors of clergy sexual abuse in Fredericksberg, Maryland, formed an organization called Healing Voices to provide low-cost therapy, retreats, and other support to clergy sex abuse survivors. What happened in the end was a surprise to the healers as they, too, found healing in the comfort of the groups. "After two years of sitting with mostly female survivors, we discovered that we, too, were feeling different about ourselves as Catholic women," reports Mary Liz Austin, a

Catholic pastoral minister and founder of Healing Voices. "So much of what we were hearing from women resonated with our own experiences."

Austin observed that both female survivors of sexual abuse and female survivors of the Church's debilitating sexual teachings share a deep wound, condemned as Catholic girls to silence and shame for their suspect sexuality, not just by priests but by judgmental nuns, too. Those teachings and that treatment compounded the experience of sexual abuse for Catholic girls, particularly older girls, who Austin has observed can take decades to admit their exploitation or abuse because they feel sure they were responsible. "We started to see how little we have provided nurture for one another as women in a patriarchal Church," Austin says.

Healing Voices attempts to provide that nurture. In addition, the organization attempts—as do other reform groups—to foster healing between the survivors and the hierarchy, who have been so alienated by the sex abuse crisis. In 2004, following a year of dialogue, Healing Voices and representatives from the Baltimore archdiocese cosponsored what observers and participants alike described as a moving atonement service. While the service was led by survivors, some fifteen to twenty priests and deacons, as well as Baltimore Cardinal William Keeler, participated. Taking a very humble role, the cardinal publicly apologized for the failings of the Church in protecting children from abuse, at one point inviting the other clerics to kneel with him before the altar, where together they recited the Act of Contrition.

The Crisis Continues

Certainly, the Catholic Church hierarchy is to be commended for some of the steps they have taken to address the sex abuse crisis of 2002, in keeping with the pope's unqualified response that there was "no place in the priesthood or religious life for those who would harm the young." They opened their Dallas meeting in 2002 to the world and their critics; they set up a National Lay Review Board with the power of voice (though admittedly a voice the bishops could cut off at any moment, and from which they excluded the leading survivor advocates); they established a national Office of Child and Youth Protection and conducted compliance audits two years running; and they authorized the conduct and public distribution of surveys on the incidence, causes, and context for child sex abuse among Catholic clergy, which even with their limitations are far more than most other institutions have ever done.

The bishops' harshest critics, the survivor activists, acknowledge some progress, too. SNAP director, David Clohessy, reports that it is definitely easier now than before to get an offending priest removed from the ministry and to get a survivor's counseling paid for. Blaine commends the bishops for establishing the National Lay Review Board, while acknowledging its limitations as a creation of the bishops' conference. Both she and Clohessy stress that where offending priests have been removed from the ministry, children are safer.

But they don't credit the bishops for the progress. "The only changes they've made didn't come from them but from outside sources, either public pressure or political pressure or financial pressure," says Blaine. "There has never been the effort to determine how to best meet the needs of victims and reach out." While the bishops did invite survivors to speak at the 2002 Dallas meeting—including SNAP's director Clohessy, because, according to Blaine, the bishops "made it clear" she would not be allowed to speak—they have not responded to any invitation from SNAP to talk with them since. This raises great skepticism in Blaine about the bishops' 2002 commitment to "the beginning of a dialogue." "They met with us when the cameras were rolling," she says. "It was more of a show."

Indeed, the crisis is far from over, and the limitations of the Church's response are glaring. The Catholic hierarchy could have taken this horrendous opportunity to really clean house. It could have hired independent investigators given the charge of reviewing all sex abuse cases ever reported to a diocese, examining all files, evaluating how each case was handled, and reporting on the disposition of each case, without depending on the bishops to volunteer that information. The hierarchy could have insisted that each diocese name the names of those accused, including all the transfers done to protect abusers and their current whereabouts, thereby clearing the names of those found innocent and validating substantiated claims.

Publicly naming names to validate claims is critical, say experts and advocates alike, for the healing and even the safety of victims, who can bear the brunt of parishioners' anger if claims are left unresolved; for the healing of a parish, whose members can become deeply divided over unsubstantiated allegations; and for the healing of other survivors, who may find the courage to come forward. Even the National Lay Review Board recommended that each diocese publish the data they provided to John Jay College on the incidence of abuse and discuss survey results with parishioners, while board member and attorney Robert S. Bennett specifically advised that the dioceses name names. It seemed that this might come to pass when Baltimore Cardinal William Keeler courageously posted the names of priests credibly accused of child sex abuse on the diocesan website—which led scores of survivors to come forward—but in the end, only a handful of dioceses released names. For complete investigations into diocesan scandals, Catholics still have to depend on judges, attorney general reports, dogged journalists, and advocacy groups like Survivors First, which has published on its website the names of priests accused and the status of cases against them.

The hierarchy could have called for dioceses to report every case of child sexual abuse to civil authorities, especially since foot-dragging Church officials enabled criminal priests to run out the clock on statutes of limitations, thereby precluding victims from bringing their abusers to justice. As a result, activist Anne Barrett Doyle points out, "the Catholic Church is sitting on the largest list of unregistered sex offenders probably in our country." Indeed, Kathleen McChesney's recommendation was that each diocese should track priests they've let go and take responsibility for knowing their whereabouts and notifying others, including other dioceses and civil authorities.

While the bishops advocated "zero tolerance"—that is, if an act of sexual abuse of a minor is "admitted or established" through an investigative process, the priest must be permanently removed from the ministry—in fact, some flexibility exists. According to Kathleen McChesney, with priests never found guilty in a criminal or civil court or who deny the abuse, the diocese's local lay boards review the allegations, assess the evidence, conclude if the incident merits removal from the ministry, and pass their recommendation on to the bishop. It was likely that this loophole enabled Bridgeport Bishop Lori to keep Monsignor Ryan in the ministry. The Vatican supports easing off the one-strike-you're-out rule; it issued a statement in 2004 declaring "zero tolerance" for priest sex abusers to be too extreme a measure.

In its report, the National Lay Review Board insisted that the charter requires any priest credibly accused of abusing a minor to be laicized or defrocked, the ultimate removal from the priesthood. In 2004, some seven hundred priests had been removed from the ministry but all indications were that very few had been laicized. Cases continue to be sent to the Vatican—some four hundred as of 2004, according to Kathleen McChesney—where they are "backed up very badly," instead of to the regional tribunals that were supposed to handle these cases more expeditiously. In fact, in 2004, a spokesman for the U.S. Bishops Conference told me there were "no plans for regional tribunals," and he did not think anyone knew how many priests had been laicized because there was no system in place to track that.

Despite the bishops' insistence that they will cooperate with civil authorities, in spring 2004, even the president of the U.S. Bishops Conference, Wilton Gregory of Belleville, Illinois, was found in contempt of court and fined for refusing to turn over the the mental health records of a Belleville diocese priest, Father Raymond Kownacki. The priest is currently being sued, along with the Belleville diocese, by a man claiming he was sexually abused by Kownacki when he was a boy; another case against Kownacki for similar allegations has since been filed. Kownacki was removed from the ministry in 1995, following allegations resulting in litigation, after he had repeatedly raped, beaten, and induced an abortion in a teenage girl who worked as his rectory housekeeper in the 1970s (her case was dismissed for exceeding the statute of limitations). Gregory did provide over four hundred pages of documents on Kownacki as part of the recent legal action, but he resisted the court order to turn over mental health records citing the priest's privacy rights. The Belleville diocese is appealing the ruling on the records. In 2003, Rockford, Illinois, Bishop Thomas G. Doran was held in contempt of court for refusing to turn over records in the criminal trial of a Father Mark Campobello, who later confessed to sexually molesting two teenage girls; the girls were molested in 1999 and 2000. That case was not ancient history, as Church officials insist of the thousands of other cases.

In 2004, it became clear that the days of Catholic Church officials shielding sex abusers by moving them around from country to country—as was done with the priests who sexually exploited Rita Milla—were not over, either. After a yearlong investigation of more than two hundred cases of children abused by Catholic priests, the *Dallas Morning News* reported that Catholic priests accused of sexually abusing children were hiding abroad—in Africa and Latin America, Europe and Asia—and working in church ministries, with the help of church officials. Nearly half of the two hundred cases involved clergy who attempted to elude law enforcement, about thirty of whom were free in one country while facing criminal investigations, arrest warrants, and even convictions in another.

Despite "overwhelming evidence" that Cardinal Law and his senior managers had knowingly exposed "substantial numbers of children" to sexual abuse by "substantial numbers of priests," Massachusetts Attorney General Thomas F. Reilly's investigation found no grounds for criminal prosecution. As a result, the Cardinal never faced legal punishment. But he also apparently never faced Church punishment, either. About a year and a half after his resignation from head of the archdiocese of Boston, Law was given safe haven in Rome as well as his own basilica to head up—one of Rome's "four most important basilicas," a Vatican official told the press.

The response of many clerics to survivors and their supporters have been woefully inadequate, too. Survivors and advocates told me about writing to bishops in an attempt to set up meetings and getting no response at all; trying to hand-deliver letters to bishops and being accused of trespassing on church property; being closed out of membership on local lay boards charged with reviewing sexual abuse allegations; and having priests to whom they and their families were very close to stop speaking to them when they became vocal advocates for survivors. Some VOTF members reported having great success working with local priests to reach out to survivors, but others reported being frustrated at the difficulty in rallying ordinary Catholics to participate; at the resistance of priests and pastors to helping to organize or to attend healing services; and at failing to even get prayers for survivors included in the Sunday Mass petitions.

It is not only the advocates who have observed the failure of the hierarchy to reach out to survivors. "Even today, some bishops and priests fail to address the issue of clerical sexual abuse in a sufficiently open manner. . . . [They] still shy away from the subject and revert to defensive postures," wrote the National Lay Review Board. "Many Church leaders refused to meet with victim support groups," they added. "Although some members of victim support groups are not always fair to the bishops and are unwilling to give credit when it is due, disregarding these groups is shortsighted and contributes to the perception of a closed and secretive Church. . . . The review board believes that the proper response is rapprochement and reconciliation, not to interpose rejection or condemnation."

Finally, while the current crisis has spotlighted the abuse of minors by Catholic priests, the reality is that the abuse of children is just the tip of the iceberg—not only in the United States but worldwide. There is no question but that the abuse of children is the most reprehensible of clerical sexual crimes. But children are not, nor have they ever been, Catholic clergy's only victims. In fact, the sexual involvement of Catholic priests with adults far exceeds their involvement with minors, and women caught in those relationships are arguably in the greatest danger of exploitation. Indeed, while Rita Milla was sexually abused as a minor, she was also shamelessly sexually exploited as an adult.

The Church's power today rests exclusively in the hands of an all-male clerical club, whose hold on power has depended on a black wall of silence. To even admit the extent of sexual involvement with women challenges the male celibate's elevated and revered state, thereby compromising not only the wall but the Church's very governing structure. That is why the hierarchy labors furiously to keep the extent of clergy sexual involvement with women a secret. But on that front, too, the clock may be running out.

5

✠

WOMEN, PRIESTS, AND THE MYTH OF CELIBACY

The issue of Catholic priests sexually exploiting and abusing adult women broke nationwide in 2001, a year before the pedophilia crisis hit the news. At a press conference in Manhattan, representatives from an international coalition of 146 religious, human rights, and women's organizations met to launch a Call to Accountability Campaign. Its goal: to pressure the Vatican to end sexual violence against nuns and other women by Catholic priests worldwide.

The international campaign was inspired by a shocking story broken by the *National Catholic Reporter*, which analyzed and published internal Church reports written by two Catholic nuns—Sister Maura O'Donohue, a Medical Missionary of Mary, a doctor, and the AIDS coordinator for the Catholic Fund for Overseas Development; and Sister Marie McDonald, member of the Missionaries of Our Lady of Africa. The reports documented the sexual exploitation of nuns by priests in twenty-three countries on five continents. While much of the reporting was done on conditions in Africa, the reports also alleged abuses in Brazil, India, Ireland, Italy, the Philippines, Colombia, and the United States. The reports were not new;

they had been reviewed in meetings with the Vatican and religious authorities as far back as 1994.

One of the most stunning allegations concerned a nun impregnated by a priest, who forced her to have an abortion that killed her, then officiated at her funeral. Priests were alleged to have raped young nuns who approached them for the required certificates to enter religious orders; to have told nuns that oral contraceptives would protect them from AIDS; to have convinced young nuns that celibacy just meant not getting married; and to have used nuns as "safe" alternatives to prostitutes in countries plagued by AIDS— with some priests going so far as to demand that heads of convents make the nuns sexually available to them.

According to Sister Maura O'Donohue's report, when a nun became pregnant as a result of sexual abuse, she was expelled from her religious community while the priest was allowed to continue his ministry. As a result, wrote O'Donohue, some of those women "were forced into becoming a second or third wife in a family, because of lost status in the local culture," or, alternately, prostitutes. As long ago as 1990, she wrote, Catholic medical professionals in Catholic hospitals reported being pressured by priests to procure abortions for religious sisters.

O'Donohue also reported that in some countries, the parishioners were up in arms, "challenging their pastors because of their relations with women and young girls generally. . . . Some priests are known to have relationships with several women, and also to have children from more than one liaison." In one country she visited she was told that "the presbytery in a particular parish was attacked by parishioners armed with guns because they were angry with the priests because of their abuse of power and the betrayal of trust, which their actions and lifestyles reflected."

At the Manhattan press conference, Mary John Mananzan—a slightly built Benedictine nun from the Philippines, a forty-year veteran of the convent who wears a traditional blue habit and is president of Saint Scholastica College—told of similar outrages in the Philippines. One involved a priest who raped and impregnated a woman, who had the child, a girl. Twenty years later, when the daughter came to the priest, her father, for help, he demanded sex in exchange for scholarship money. Mananzan accompanied the daughter to file a lawsuit, but when the time came to serve the papers, the priest had disappeared. With another priest known to be a sex abuser, Mananzan and other nuns in her order complained to the bishop, who said it was between the nuns and the offending priest; the nuns subsequently refused to take Communion from him. While the Church is "quick in censoring nuns going to conferences," said Mananzan, referring to the Vatican's chastisement of Sister Joan Chittister, "they do not exert the same zeal in dealing with truly scandalous behavior."

Yvonne Maes, a nun for thirty-seven years in Montreal, Canada, spoke at the press conference of becoming a missionary to Lesotho, South Africa. While attending a Church-sponsored retreat, she alleges she was raped by her spiritual mentor, an Irish priest, and subsequently became trapped in an abusive relationship. The Church, she reported, took no action, except to try to silence her. She eventually left the convent and her order to tell the whole story in her memoir, *The Cannibal's Wife*.

"We want to hear publicly what the Vatican is doing to correct this pervasive and outrageous problem," said Maureen Fiedler, a nun with the Sisters of Loretto, a board member of Catholics Speak Out, and a longtime advocate for women's reproductive and sexual rights in the Church. Of the coalition's demands, she said: "We want to see cooperation with civil authorities in the prosecution of those who are committing what are in all global societies,' crimes. We want to see reparations for the victims of sexual abuse. We want nuns reinstated in their communities. And we would like to see a public apology."

After the press conference, about one hundred protestors—wearing white and purple, the colors of women's suffrage—marched through midtown Manhattan from the UN headquarters to the Vatican's Permanent Mission to the United Nations for a rally. They were delivering a petition addressed to Pope John Paul II, which called on the Vatican to punish the priests and make reparations to the nuns. Earlier, the Holy See's representative at the United Nations, Archbishop Renato Martino, had refused to meet with the coalition's leaders. In a gesture symbolic of women's place in the Church, Catholic theologian Mary Hunt, cofounder of the Women's Alliance for Theology, Ethics, and Ritual, dropped the oversized manila envelope containing the petition between the black metal bars of the mission's locked gate.

In his letter declining the coalition's request for a meeting, Archbishop Martino wrote back asking the members to "please join me . . . in continuing to pray for our priests." "We are so invisible that it doesn't even occur to them that there's something wrong in a letter about the sexual abuse of nuns by priests not to mention praying for the women," said Frances Kissling, the president of Catholics for a Free Choice, principal organizer of the Call to Accountability Campaign.

Shortly after NCR's article and its publication of the reports, the Vatican acknowledged the abuse while attempting in the same breath, and characteristically, to minimize the suffering of victims and deflect attention. "The problem is known, and is restricted to a geographically limited area," said a mind-boggling Vatican statement. It also said that the Holy See was "dealing with the question in collaboration" with other Church leaders, and would be focusing on the general area of seminary training and "the solution of single cases."

The Truth About Celibacy

Statistics show that true celibacy for the entire Roman Catholic priesthood is a myth. An estimated half of all priests are involved in some kind of sexual activity at any one time, according to the research of psychotherapist and former Catholic monk Richard Sipe: 30 percent are involved with women, 15 percent are involved with men, while just 6 percent are involved with minors. The National Lay Review Board made a similar observation, though it did not differentiate by gender. They reported having heard from "numerous witnesses" that there were "more incidents of sexual relationships between a priest and a consenting adult woman or man than between a priest and a minor." They characterized the women and men involved as "often vulnerable" and the priest's behavior as "gravely immoral." According to a recent Los Angeles Times poll, only a third of priests in the United States say that celibacy is not a problem for them.

Despite the level of sexual activity that exists among Catholic priests, the Church has blatantly and belligerently refused to deal seriously with breaches in celibacy. "Celibate law is extolled while celibate transgression is indulged, fostered, easily forgiven, and covered up," says Richard Sipe. "Celibate transparency and accountability are nonexistent in the Church, and attempts to foster honesty are labeled disloyalty, impudence, or heresy." It is his contention that "the inherent duplicity between the stated norm, belief, and practice thrives on the denial of sexual reality," and that "this communal dishonesty sets the stage for sexual corruption and abuse."

The Burden of Duplicity

Many adult Catholic women have borne the burden of this duplicity. As victims of the priest's unique spiritual power and privilege, they have been subject to clerical transgressions from sexual exploitation to harassment to molestation to rape to beatings to potentially negligent homicide. Many sexually active priests have left a trail of wounded women and fatherless progeny in their wake—terrible testaments to the hypocrisy in the claim of a celibate priesthood.

In 2003, plaintiffs' attorney Roderick MacLeish, Jr. succeeded in getting the court-ordered release of more than 45,000 pages of documents pertaining to 141 different priests in the archdiocese of Boston who allegedly sexually abused minors. Some of those same priests had blatantly abusive involvements with adult women as well, which were handled by diocesan officials with the same callousness toward victims and commitment to coverup as they had handled cases of child sexual abuse.

One such case involved Rita J. Perry, covered in detail by the *Boston Globe*. In 1973, Perry took an overdose of sleeping pills. It wasn't the usual scene; however, she wasn't alone. In the house that night was a priest who had began a sexual relationship with her when she went to him for counseling. A philanderer who came in and out of her life, the priest had fathered two of her children. In the house that night, too, was her infant daughter, in a crib. Instead of taking Perry to the hospital, just a few miles away, the Reverend James Foley fled the scene. He returned later to find her dead, and only then called the police—anonymously.

That was the first story Foley confessed to Cardinal Law and his aide Reverend John B. McCormack (now bishop of Manchester, New Hampshire) in 1993. Foley later modified his story to say that he made the anonymous call for help *first* and fled later. Bishop McCormack did not notify the family or the police when Foley told him what had happened. But McCormack was quite cognizant of the potential legal responsibility. In a note to himself that he put in Foley's file, McCormack wrote: "Criminal activity?" Foley managed to remain in active parish ministry until December 5, 2002 when his file became public. Amazingly, the publicity led his two children to suspect Foley was their father, a suspicion confirmed by a court-ordered DNA test; in 2004, the Boston Archdiocese settled a wrongful death suit with the family.

Another cleric, Reverend Thomas P. Forry, was accused of viciously beating up his fifty-eight-year-old housekeeper twice, in the 1970s, trying to throw her down the stairs and literally pulling "a large amount of hair forcefully . . . out by the roots." Instead of sending Forry to the authorities, the archdiocese assigned him to a new parish. Later, another woman came forward. She claimed that Forry had been acting as her "husband" for eleven years. He had built her a house on Cape Cod but then threw her out of the house and left her destitute—but not before sexually and physically assaulting her son.

Despite this horrendous history, Cardinal Law selected Forry in 1988 to serve as a U.S. Army chaplain. In his assignment letter to Forry, Law gushed: "I have every confidence that you will render fine priestly service to the people who come under your care." Afterward, Forry was put on temporary parish assignments to fill in for absent priests, where he finally met his demise. In 2002, a parishioner reported that Forry had sexually molested another boy and his sister, and Forry was finally removed from the ministry.

Auxiliary New York City bishop the Reverend James F. McCarthy resigned in 2002 after admitting to "a number of affairs with women over several years," including a woman "approximately twenty-one years old." McCarthy had worked for two of the most conservative bishops on sexual matters, powerful heads of the New York archdiocese—Cardinal Edward Egan and the late Cardinal John J. O'Connor. As for U.S. bishops in general, by 2004, at least fourteen had been accused of personal sexual misconduct, with more than one-third of those cases involving adults.

Susan Archibald, president of The Linkup, the nation's second-largest group working with survivors of clergy abuse, has heard plenty of stories from adult women sexually exploited by men, or worse, by Catholic priests. Archibald herself is an adult survivor. As a freshman cadet at the U.S. Air Force Academy, she was sexually exploited by the priest chaplain, a captain to whom she was ordered to go to for counseling at a very difficult time in her life. Though she had "a mountain of evidence" and proof of two other victims, the military took no action. "They let him sit there because they didn't want to bring a scandal to the Air Force Academy," she reports. "The day I gave testimony, the colonel who was in charge said there wouldn't be a court martial because it would be a publicity nightmare." Not until six months later, when she told her story to the *Denver Post*, was the priest relieved of his duties.

Since she began serving as director of The Linkup in 2000 and has told her story, Archibald says she has been contacted by 450 women saying that they, too, were sexually abused as adults and could get no action from the Church. "It's astounding to find out how prevalent it is," she says. And while the abuse is bad enough, she adds that when women bring their cases forward, they encounter a whole other layer of risk. "There's a real sense of fear," she says. They come to see that they will "be examined or blamed and embarrassed" because of what happened to them. In some cases, she adds, "especially with adult victims, when it gets around the parish, they get physically run out of town."

That is exactly what happened to Judy Ellis, sexually exploited by a Catholic priest from 1996 to 2000—one of the women who turned to Archibald for help. When her marriage was falling apart, Ellis went to see Father Abraham Matthews, then priest at Saint Catherine of Sienna Church in Atlanta, Texas. Matthews quickly turned their counseling sessions into opportunities for sexual exploitation, Ellis told me, which soon became drunken scenes of escalating physical violence, against her as well as her daughter. The destructive relationship lasted for five years, until the diocese transferred the priest back to his home parish in Africa.

That was when Ellis began to put her life back together, which included filing a lawsuit against the Diocese of Tyler, Texas, and Matthews for sexual exploitation. Unfortunately, because parishes virtually never confirm the sexual exploitation of adult women by priests, when the parishioners found out about the lawsuit and the priest's disappearance, Ellis became a pariah. "They slashed my tires and tried to burn a cross on my lawn," she reports. Eventually, unable to take it anymore, she and her daughter moved with the help of a woman friend to another state. The diocese settled with her for an undisclosed sum, which she is forbidden by a confidentiality clause to reveal.

While men, too, are victims of sexual exploitation by priests and contact The Linkup, the vast majority—95 percent—of those Archibald hears from are women. When the exploitation begins, the women range in age from eighteen to over sixty and are in the throes of a life crisis, most commonly a marital crisis. They go to priests seeking advice about how to work on the relationship, explained Archibald, and "the priest . . . convinces the woman that [he is] the perfect solution, this is what God wanted, goes on to exploit them, then moves on to someone else."

Archibald says that almost all the women she's talked to have been suicidal at some point. "Some have been hospitalized. Some become alcoholics, get involved with drugs, a lot of the same effects and impact you see with the child victims." To her mind, it is the blaming of the woman that has the worst results. "The last thing you need when you go to tell the story to someone in the Church is they turn around and blame you for causing poor Father to sin," she says. "There's no recognition [of] or compassion [for] the effect it's had on your life. So that betrayal in itself drives women to the brink." Psychologist and expert in the area of clerical sexual abuse Gary Schoener adds that strong women get involved, too. For them, he says, "it's normally the timing. She's at a low ebb in her life or very needy or something unique has occurred." Or, he adds, she may be a very bright, thoughtful, and spiritual woman who develops a "a unique connection to this priest," who becomes a "key part of her life . . . a partner. Plus, he's a direct link to God." As to the appeal of these priest exploiters, Schoener notes that some are "very bright, stimulating, interesting. . . . It's important to get away from the notion that all these priests are losers; they're some of the more impressive people you ever want to meet."

If teenage girls are blamed for a priest's sexual transgression, adult women don't stand a chance. They are considered equally culpable. Yet the behavior of the man with power is where the blame lies, as described over a decade ago by psychiatrist Peter Rutter in his groundbreaking book *Sex in the Forbidden Zone: When Men in Power—Therapists, Doctors, Clergy, Teachers, and Others—Betray Women's Trust.* His contention is that any sexual behavior by a person in power inherently exploits the trust of the person without the power. He writes that it is the responsibility of the man with power "no matter what the level of provocation or apparent consent by the woman, to assure that sexual behavior does not take place."

To Schoener and many other mental health professionals, a violation in the process of spiritual counseling is the worst violation of all. In addition to using their spiritual power "as agents of God" and their unparalleled "aura of safety," compliments of celibacy, to gain trust and compliance, Catholic priests—like all clergy—have enormous access. Says Schoener: "All clergy . . . are on duty twenty-four hours a day. . . . They can meet someone anywhere. They can meet with you in your home. If you're in your bedroom sick, they can come up. . . . They can tuck your kids in and say prayers by their bed." Priests can and do counsel women in cars, on drives to meetings, in their offices, the woman's living room, and out on "dates," leaving their collars at home. "The more religious the victim is," adds Schoener, "the more tightly [he or she adheres] to the faith, the easier targets they are. The more the pastor or priest claims divine inspiration, the better they are able to exploit."

Incorporating the principle that real sexual consent is impossible in the inherently unequal power relationship between a helping professional and a client, some twenty-three states have laws making it a crime (in almost all of these states, a felony) for a psychotherapist to have sex with a client. "In most of the states, if a clergyperson is doing psychotherapeutic counseling . . . [for] an emotional or mental health problem, they are covered," Schoener writes. And in most states, he adds, consent is not a defense, though some require proving other facts, such as that the sex was the result of emotional dependency. In two states, Minnesota and Texas (where Judy Ellis brought charges), the laws go even further, explicitly covering clergy doing spiritual counseling.

About a dozen of the women Susan Archibald has heard from have borne children. Some of the priests are involved in the children's lives; some never see them again; and some condemn them to a kind of limbo. Archibald tells of a woman in Alabama whose child "would hang around in the back of the church, then come up after Mass and say, 'Hey, Daddy!' I guess it was understood, some people knew about it, but nobody wanted to recognize that as being a problem." A lot of women involved with priests have tried to sue the priest's diocese for child support, she says, but the dioceses maintain that it is not their problem.

Archibald, as well as Schoener, has worked with women encouraged by their priest abusers to have abortions. Psychologist Richard Sipe, too, has observed dozens of cases. Indeed, I spoke to a woman who was sexually exploited by a highly placed Catholic cleric, a dear family friend, when she was twenty-one years old and single, and he was more than twice her age. Aware that she was deeply depressed and distraught over recent losses in her life, the priest offered to counsel her, which was a great comfort to her and her family. Over time, the priest turned their association into a sexual one. She became pregnant twice, and twice he took her out of the country and paid for her abortions. As she tried over a period of years to pull away, he began to harass her at her job, threaten her, and stalk her. The woman married, and it was her husband who put a stop to the torment. But the woman spent years in guilt and shame, until she was able through the survivors' movement to forgive herself and help other sexually abused women to heal. She cannot, however, bring herself to talk publicly about the abortions.

On the subject of children, Archibald adds that she knows of cases where priests who exploited adults were allowed to remain in the ministry, only later to be accused of child sexual abuse. By ignoring the priests' transgressions with adult women and not removing those priests from the ministry, the Church put children at risk, too.

Secret Lives, Broken Vows

But that is really only half the story. In fact, celibacy as it pertains to women is a complex issue. Clerical sexual relationships with adults exist on a continuum, from the abusive to the consensual. Not all sexual involvements between priests and women qualify as exploitative. They may not involve a counseling relationship or the exercise of spiritual power; the woman may live in another diocese, having nothing to do with the priest's church or parish activities; or they may be work colleagues. Some of these involvements have resulted in successful long-term unions by those who left the priesthood—an estimated twenty thousand since the 1960s—most to wed.

Such was the case with Cait Finnegan and her husband, Joe Grenier. Back in the 1970s, after leaving the convent as a young novice, Finnegan fell in love with Grenier, then a priest. She reports that he "prayed his way through" the very difficult decision about whether or not to leave the priesthood. That took two years, during which the two remained celibate, waiting, to her dismay, "until the day they were married" to have sex.

As evidence of the hopefulness with which they started out, Finnegan and Grenier began in 1983 to build an organization called Good Tidings. Their goal for the organization (originally founded by a woman with a priest lover who left early on) was to help couples like themselves. Finnegan confesses, "When we started, our presumption was that these relationships [consisted of] friends who had grown to love each other and needed to discern what God was asking," that is, for a recommitment to celibacy or a decision to leave to marry. She says they found hundreds of couples who did not want to live a double life, for whom making the decision to leave was a matter of the utmost integrity. Those priests, she reports, "became members of CORPUS (which advocates for a married and single, male and female priesthood), as well as primary donors to Good Tidings."

But Finnegan has worked—and continues to work—with couples where the priest remains in ministry while maintaining a secret, committed relationship with a woman. Some of those couples have had civil or legal weddings or at a minimum, taken private vows. "Some of them," she says, "are raising families." Indeed, according to the research of Richard Sipe, of the 30 percent of priests involved with women, two-thirds are in either "a more or less stable sexual relationship" or in relationships with "sequential" women, while the other third are having "incidental sexual contacts." He does not estimate the number in consensual versus exploitative relationships. Sipe, too, found priests who had entered into civil marriages and continued their ministries, as well as priests having children, noting that the

Church and their social service agencies at times fought the opening of adoption records for just that reason—to keep secret a priest's paternity.

From what I have observed, the subject of Catholic clerics involved in long-term relationships with women is fraught with controversy in reform circles. Some Catholic reformers believe that the priests who have solid relationships are probably "the healthier ones." I remember sitting with a group of Catholic reformers in the early 1990s talking about a bishop everyone knew had a girlfriend. They refused to call him on that. He was not the enemy, they said. They had nothing to gain by outing him. To me, in a Church whose clerics regularly denounce the sexual practices of so many ordinary Catholics, who have caused those ordinary Catholics—especially women and girls—so much guilt and pain, that smacked of the worst hypocrisy. So when Finnegan found me a woman in a long-term relationship with an active Roman Catholic priest who was willing to talk to me, I was eager to hear what she had to say.

Marilyn (not her real name) has been in a committed, sexual relationship with Father Tom (not his real name) for twenty-four years. She turned to Good Tidings early on for advice and counsel. Marilyn met Father Tom when she was converting to Catholicism. She was twenty-eight at the time; he was three or four years older. "There was a real spark there," Marilyn tells me, with obvious delight, explaining the early days of their relationship just as any woman still in love with her partner might describe it. I ask for details. She tells me they first slept together after a dinner date and a night of dancing. She went home with him to the rectory, leaving—as she describes it, nonplussed—at about 5:00 A.M. to avoid a face-off with the crowd coming in for six o'clock Mass. Next, they got to "spend a whole day and night together," ending with dinner and the privacy of a hotel room. The relationship grew from there.

Marilyn says, "I never looked at him as a priest. I don't look at people's vocations for who they are." She also never had any intention of trying to get him to leave the priesthood. In fact, it makes her crazy when women involved with priests have that as their goal. "Say you fall in love with an obstetrician," she explains. "Do you tell the obstetrician, 'Look, I want you home every night'? Guess what? Babies come 24/7, 365. And it's the same thing with a guy who's a priest." Well, not exactly. It is the first of her creative rationalizations for the controversial position that she is actually in.

Marilyn tells me that her not wanting children or to live "24/7" with someone were factors that helped sustain the relationship. But after several months of involvement with Father Tom, she realized what she did want: to take the relationship out of the closet, at least with family and close friends. They did. He began to take her to Sunday dinner at his parents' house, she introduced him to her family, and they developed a circle of friends. Marilyn believes the reason they have survived as a couple is that "we have the support and love of our families and friends. We chose not to live in the darkness where many women and men find themselves—nothing can grow and thrive in darkness."

Beyond their intimate circle, other people know they are "great friends" and that they often "travel together." A few priests have known about the relationship, but Marilyn is unsure what the local bishop knows. Her impression—and she is an insider in the diocese, being employed in a parish rectory—is that it isn't something he would make a big deal out of. Word on the street is that "he wants no scandals," and that it's okay as long as it's not public. "The only way he's going to complain," she jokes, in what I come to see is her ribald fashion, "is if he has to walk over you to get to the cathedral on Sunday morning and you're screwing on the steps."

Father Tom has worked in priest recruitment and as a parish priest—though never in the parish where Marilyn works, nor in the parish where she worships. She insists that her worship parish be different from the parish where he's employed—not because the relationship must be hidden but because people don't follow their partners to work. Another creative rationalization.

When asked about Father Tom's ministry, Marilyn talks about the last parish that Father Tom took over, which she describes as having been "really wounded" when he arrived. Ironically, one of his two immediate predecessors was a child abuser who had to be removed, while the other had had an affair with a young married mother, a parishioner, which became known to the parish community. I have to say that made Father Tom look pretty moral to me by comparison. Marilyn is proud of Father Tom's work at that parish: "He did really wonderful things," she says. "The people loved him."

It interests me, too, that Marilyn and Father Tom have crossed other forbidden lines, familiar to ordinary sexually active Catholics. Because she didn't want children, they obviously had to use some kind of birth control, and it was not the Church-approved rhythm method. They used an IUD and the pill. And Marilyn is an unabashed advocate for women's access to reproductive health care, pleased that the California Supreme Court just voted that Catholic Charities must provide insurance coverage for birth control to its female employees to whom they provide general prescription drug coverage. She is pleased because she believes that pressure from the outside is one of the ways that the Church will change.

But she also believes that change comes from inside the institution, which is why she is willing to talk to me. She wants relationships like hers to come out in the open, so that people can see that priests can have committed relationships and still be good priests. She admits that the other women on the Good Tidings's Listserv "get very angry with her," accusing her and Father Tom of hypocrisy, but she doesn't agree. "You know what I believe about Catholics?" she says. "That it really doesn't make a difference what you're doing in your personal life. . . . If you're a good man that loves doing this work, or a good woman, then all that other stuff doesn't mean a bit of difference."

Marilyn reports that she and Father Tom have survived some "rough patches," like when he had an affair, which sent them into couples counseling, which seems almost surreal to me. Here is a man who is supposed to be celibate. He breaks that vow by having a sexual relationship. Then he violates that commitment by having an affair. Then he goes into therapy to learn how to better break just one vow.

It strikes me that Father Tom and Marilyn have created an alternate universe, which many Catholics do as they struggle to behave morally while rejecting the Church's sexual straitjacket that forbids all sex outside of marriage and any sex closed to procreation. I appreciate the terrible loneliness that priests can suffer, particularly parish priests banished to rectories, where they are supposed to live alone and are disconnected from sex and intimacy. I appreciate that Father Tom is committed to Marilyn, in contrast to other priests who might abide the recommendations of their superiors, as described by Cait Finnegan, "to use her and get rid of her." I'm impressed that the two have every intention of growing old together; they just took a big step and bought a house by the beach.

On those levels, I understand the way they live, but on another more basic level, I am deeply disturbed by it. This man has great power as a priest, holding himself forth as a spiritual model of celibate fidelity and holiness, when he is not. "Personal celibacy is a public stand," writes Richard Sipe, and I agree. Furthermore, by staying silent and hidden, the couple reinforces the Church's propensity for duplicity and cover-up. They live secret lives and demand that their intimates join them in the lie so that the whole enterprise becomes an exercise in hypocrisy. That Marilyn's priest lover works in priest recruitment is doubly disconcerting, for this man who has rejected his celibacy vow is the one counseling new priests on how to keep that vow—or not. I don't think a compact among clerics to collude in denying sexual reality will change the Church for the better.

From 1983 to 2002, Cait Finnegan had in-person, phone or e-mail contact with nearly two thousand women who have been sexually involved with priests. Nearly fifty had children by the priests, some more than one; the most recent birth she knew of when we spoke was on Christmas Day in 2003. She continues to hear from women in relationships with priests who are moving toward marriage (she estimates that about 10 percent of the women she's worked with married their priest lovers), but that is not the lion's share of her work today. She estimates that three-quarters of the women who contact her are in an exploitative relationship with a Catholic priest. Finnegan helps those women to find services, counseling, and other women in the same situation who can help them break free from what can become devastatingly destructive relationships. If children are involved, Finnegan advises the woman not to go to the diocese for help but to take the priest to family court. Under the deadbeat dad laws, she advises, "Uncle Sam can demand a paternity test," and if it's positive, "Uncle Sam can demand pay for child support." She remains amazed at the extent of the problem of the exploitation of women by Catholic priests, and the degree to which it is "being totally ignored."

Finnegan herself was sexually abused as a high-school student and then as a novice by a nun. She is concerned that this type of abuse, which she calls "a very silent reality," has received so little mention in all the talk of the sex abuse scandals in the Church. While Good Tidings does not deal with women abused by nuns, Finnegan and author Ashley Hill (who wrote about her experience of abuse in her book, *Habits of Sin*), formed a support group for victims of sexual abuse by nuns. As for the institution's handling of her case, Finnegan is satisfied. Representatives of the Sisters of Mercy convent to which she and her alleged abuser belonged have met with Finnegan in her home, are paying for her counseling, and removed the nun from the ministry—all this as Finnegan commenced a lawsuit against them.

Women's Labors Lost

Catholic women in the ministry have attempted to rescue the Church from the jeopardy of clerical sexual indiscretions, but at a cost. One campus minister—who chose to remain anonymous—was part of a Midwest diocese in the early 1990s. They had in-service training to heighten awareness of sexual exploitation by clergy and to encourage reporting such incidents internally. Following that, the campus minister learned that a priest on her team was involved with a vulnerable young man, a parishioner.

First, she refused to lie about it. "When you're on a team and you know your priest is having an affair with a parishioner, you have to lie. You have to lie to parishioners. . . . You're put in the position of all of a sudden being torn between what you know in your gut is right and worrying about keeping your job." And she knew the parishioner, who she saw "was really struggling" as a result of his involvement with the priest. For his sake, she felt compelled to act. But she also believed that reporting the priest was a way to head off an unhappy ending for the diocese, as "the right thing to do to protect us."

"Do you want to know what they told me when I went in to report Father So-and-So?" she asks. "My director told me, '[Celibacy] is a promise. It's not a vow,'" which is true. But that's not all. She also reports that he said:

"You dated a parishioner. What's wrong with Father So-and-So dating a parishioner?" Given the clear difference between her job as campus minister and a priest who served Mass and had enormous power in the parish, and his commitment to celibacy versus her single state, she felt that was a tremendously unfair comparison. The minister reports that they put a "gag clause" on her, threatening that "if I should mention it to anybody, I'd be fired." Instead of waiting for the ax to fall, she left that position.

Sister Celine Goessl was a pastoral administrator at different Catholic parishes in the United States for nearly thirty years. She worked with many priests who were sexually active, including a pastor who lived with his male lover away from the rectory. One was an active alcoholic who spent night after night at the local bar picking up men and bringing them back to the rectory. Word got back to Goessl, who confronted that priest about it and went to the local bishop. But instead of the priest being removed, she was. "It was December twenty-third, just before Christmas," she recalls. "I went home, I holed myself up in my room at the Mother House, and didn't come out all Christmas."

Judy Diaz Molosky, a sister with the order of Saint Joseph of Carondelet in Los Angeles, was hired in September 2002 as the certified pastoral associate in charge of outreach and social justice at Saint Brendan Church in Hancock Park, a wealthy area of L.A. Molosky assisted Father Charles Gard, until he was alleged to have had affairs with female parishioners, including his secretary, who blew the whistle on him. A crisis ensued, which resulted in Gard being relieved of his duties, and Molosky being invited to assume the role of pastoral administrator. It was a huge job, at a very volatile time, but Molosky pulled the parish through, with the support of dedicated priests and the local bishop. Molosky apparently did so well helping the parishioners cope with the loss of their pastor under trying circumstances that when the seventy-two-year-old pastor of a nearby parish was "busted" for paying prostitutes, she was invited to manage that crisis as well. She declined.

For six months, Molosky managed Saint Brendan's administrative, ministerial, and spiritual life. But then, in June 2003, Molosky was unceremoniously replaced by a new priest pastor. He came in, took over, and failed to renew her contract. This is not an unusual situation. But instead of docile acceptance, Molosky and her entire religious community fought back. In an action reminiscent of the Erie Benedictines, nearly two hundred of them signed a letter to Cardinal Roger Mahoney of the archdiocese of Los Angeles to express "serious concerns regarding the misuse of authority by new pastors." Letters of protests poured in from the laity as well, some of whom withheld contributions.

To quell the storm, Cardinal Mahony transferred the new pastor out of the church. In a letter to Molosky's community, he promised to establish a special committee to work on "collaborative models of leadership" between lay ministers and clerics, which he did. But he did not reinstate Molosky to her job. In fact, she did not want reinstatement. While admitting that her time as pastoral associate at Saint Brendan's was "the most meaningful ministry of my thirty-five years of service as a sister of Saint Joseph of Carondelet," she will not even think of resuming such a position unless the Church institutes policies that respect the contributions and workplace rights of lay employees.

The Road Ahead

Both Kathleen McChesney, director of the bishops' Office of Child and Youth Protection, and the Lay Review Board argued for consistent, mandated policies to reach out to vulnerable adults sexually exploited by priests and to handle their complaints in a systematic way. Referring to breaches in celibacy, the board wrote: "Bishops and other church leaders cannot allow such conduct to occur without consequences. . . . Whether a priest keeps his vows and lives in accordance with the moral precepts of the Church is the business of his bishop, his fellow priests, and his parishioners." They characterized this issue as "a breeding ground" for another crisis.

Estimates are that some 30 percent to 50 percent of Catholic priests have a homosexual orientation, and the Board made special note of sexual activity among gay Catholic priests. While refuting the right-wing position that gay priests caused the sex abuse crisis, and commending the many gay priests who lead devoted celibate lives, they also referred to the formation within the Church of a "gay subculture," featuring "homosexual liaisons among students, between students and teachers, and within certain dioceses and orders," activity that went ignored and unaddressed by the priests' superiors. Moral theologian Sydney Callahan has her own idea about what may have contributed to that specific area of secrecy. "Within the Church's distorted sexual teaching focused on reproduction and the danger of adult women," it may be easier to see "homosexual encounters . . . as safe from consequences," she writes, and to keep those encounters "relatively invisible." The board called for "bishops, provincials, and seminary rectors to ensure that seminaries create a climate and culture conducive to chastity."

But chastity has proven to be an unattainable goal for the Roman Catholic priesthood as a whole, and Church authorities know that. Even with the crisis of sex abuse that they have just endured, there seems to be no change in their powers of denial on the subject. As evidence of this, a group of bishops chastised Kathleen McChesney in writing for daring to handle cases involving priests and adults. In reality, McChesney told me that gathering reports from adults was not part of her mandate, but such reports have come to her and when they do, she refers people to whatever help she can, including legal authorities.

In the United States, a vigorous movement for optional celibacy among Catholic priests emerged in the wake of the sex abuse crisis, which has ramifications for the Church's view of and treatment of women. That movement began in 2003 in the archdiocese of Milwaukee, when 163 priests signed a letter to Wilton Gregory, president of the U.S. Conference of Catholic Bishops, urging the opening of the priesthood "to married as well as celibate men." It made no link to the current extent of adult sexual activity among Catholic priests. Rather, it declared as its motives the priest shortage as well as "the ever growing appreciation of marriage and its many blessings so compatible with priesthood and even enhancing of priestly ministry."

By bringing women into the intimate circle of Catholic clergy as wives, optional celibacy would begin to challenge the basic misogynistic beliefs that continue to underlie the celibate priesthood. "Mandatory celibacy is built on the notion that women are inferior, and marriage is a second-rate way of being Christian," says Anthony Padavano, a former priest and longtime member of CORPUS. The opportunity to change the Church's views

of women through the acceptance of married priests, as
suffering of the women who one reformer called "the causa
sory celibacy," are major reasons that women reformers, too, a
optional celibacy movement.

As the movement spread to priests in other dioceses across the
parallel lay campaign added to the momentum, spearheaded by two na
progressive Church reform groups—Call to Action and FutureChurch
by women, Linda Pieczynski and Sister Christine Schenk, respectively.
mid-2004, the Corpus Christi Campaign for Optional Celibacy had succeeded in collecting some fifteen thousand letters for Bishop Gregory from
ordinary Catholics advocating an end to mandatory celibacy. In addition,
Church reformers in fifty-three dioceses polled their own priests; they found
that 67 percent of the 2,589 priests who responded (out of 3,846) favored
open discussion of the Church's mandatory celibacy rule.

This movement for married priests should have more impetus than ever
because in the Catholic Church today, there are *already* married priests. The
Church's Eastern rite priests (e.g., the Ukrainian Catholic Church) have
been marrying for centuries, and there are married Roman Catholic priests
who wed before they converted from various Protestant denominations,
which the Church began to allow in 1980. On the other hand, many new
seminarians are more orthodox than their predecessors and vocal advocates
for mandatory celibacy, which has been Pope John Paul II and the Vatican
hierarchy's unflappable position. Indeed, in the Call to Action/Fu-
tureChurch survey, while priests aged forty-one to seventy most strongly favored discussion of the mandatory celibacy rule (74 percent), the majority of
the priests under forty did not (54 percent).

In addition, in an effort to dissuade men from leaving the priesthood,
dispensations from marriage have been regularly refused, particularly for
men under forty, during this papacy, according to former priest Anthony Padavano. Padavano describes the process today as "very punishing," and reports that a priest is only guaranteed a dispensation if he's dying or if he is
willing to state on his application that he was not freely ordained, that the
priesthood was a danger to his salvation, or that he has a severe sexual disor-

ı leaves the priesthood to marry and
diately; if he divorces and has no fi-
iission as well.

ommittee on Sexual Abuse had
dults by Catholic priests. That
s. Survivor Susan Gallagher,
, and law at the University of
cious initiative. As a leader of the Co-
vivors, she is working with law professors to de-
will help victims and advocacy groups to document the
abuse, exploitation, and violent assault of women by Catholic
priests. They will borrow the strategies of plaintiffs' attorneys and news out-
lets, which succeeded in getting the release of Church files and previously
sealed court documents in cases of priest child sexual abuse, based on the
public's right to know. "The public has as much of a right to know about
crimes against women as it does to know about crimes against children," insists Gallagher.

Susan Archibald of The Linkup has developed a website for adult survivors, which includes a database of adult perpetrators. In the interest of beginning to document the problem, Archibald developed a basic reporting
form that she is using with calls coming into The Linkup and sharing with
Church officials. She also reports that representatives of The Linkup have
begun to work with religious orders through the Conference of Major Superiors of Men, "in a cooperative way to help educate and examine current and
future problems," including "having us provide input for policies and training on the adult issue." By mid-2004, The Linkup had completed an educational video on the abuse of adults by clergy, for training use by members of
religious orders.

But Archibald is very clear that much more needs to be done. "The
abuse of vulnerable adults by priests has yet to be sufficiently addressed by
the Roman Catholic Church or its offices," she has written. "The Church
has not recognized these behaviors as potential criminal conduct, despite
the fact that . . . states have criminalized sexual contact between clergy and
vulnerable adults. Diocesan policy must address the sexual abuse of adults

with the same seriousness as violations against children, and remove those who offend."

It is obvious that the crisis in the Church is much larger than pedophilia or child sexual abuse. It is about crimes and criminals, sex and power, yes. But fundamentally, it is about hypocrisy. By forbidding priests who choose to be sexual to do so in mature ways that include commitment, responsibility, and respect, and by protecting them from the costs of their sexual exploits, the Church has effectively condoned a clerical sexual free-for-all. That heterosexual and homosexual behavior may thrive in the Catholic priesthood does not reflect anything inherent about homosexuality or heterosexuality but is rather an indictment of the hypocrisy and duplicity of an elite, closed, all-male system, a secret society of sorts, that condones, indeed, demands, lying about the reality of one's sexual life *at all costs*. The Church requires this in the interest of maintaining the public fiction of a celibate priesthood made up of male models of spiritual living and integrity, entitled by their extraordinary virtue to absolute power.

But that denial of sexual reality does not apply only to Catholic clergy. It also applies to Catholic laity, caught between the Church's intransigent teachings about sexual intimacy and the facts of their real lives. "The magisterial teaching is that every sexual thought, word, desire, and action outside of marriage is mortally sinful," explains Richard Sipe. "Every sexual act within marriage not open to reproduction is also mortally sinful." To his mind, this "aberrant teaching" is not a "reasonable norm for developing boys and girls or mature adults . . . who cannot, and do not, live or grow this way."

One might have expected, in the face of the onslaught of outrage engendered by the child sex abuse crisis, that the hierarchy would back off its rigid prohibitions concerning premarital sex, artificial birth control, infertility treatments, condoms to prevent the spread of AIDS, sterilization, homosexuality, remarriage after divorce, and pregnancy termination. But that has not been the case. In an effort to regain its moral footing in the years following the sex abuse crisis, the Catholic hierarchy labored tirelessly to switch the focus from the sins of the fathers to the sins of ordinary Catholics.

Despite all that has happened, a hierarchy so obviously unable to handle its own sexuality continues to sit in arrogant judgment of the sexuality of others. There are Catholic women fighting hard to protect people from crumbling under the weight of the Church's unconscionable self-righteousness. In addition to women leading the battle to call the Church to accountability on clerical sexual abuse, women are working hard to defend women's reproductive health and moral authority; to minister to the Church's sexual outcasts, from homosexuals to single heterosexuals to the divorced and remarried; and to create a new Catholic sexual ethic. Their battle is tough and generally thankless. But for the people and for the Church, it is one of the most urgent battles of all.

6

✠

THE "A" WORD, THE "C" WORD, AND THE FIGHT FOR WOMEN'S MORAL AUTHORITY

The afternoon I visit Frances Kissling, she is just hours back from a trip to China. You'd never know it. I've known Kissling for nearly twenty years, and I've never seen her look bedraggled. It's September, and she's padding barefoot around her bright Washington, D.C., apartment. She's wearing a black V-neck jersey, white knit capris, and no makeup on her clear, pale skin. Her short, bobbed, blond hair is stylishly unkempt, and her blue-gray eyes are as clear as ever.

Kissling is a blend of a pixie and a snapping turtle, in the swiftness of her tongue and, if she doesn't like you, her bite. Her mind is as big as all outdoors. She loves to laugh, and does so with gusto. And she's fearless. In part, that's because Kissling has located herself at once inside the Catholic Church but also outside of its draconian disciplinary reach.

When I visit, there has been an outcry worldwide over the decision of the Bush White House to once again withhold all monies from the United Nations Population Fund. UNFPA is the world's largest international funder of reproductive health programs. It supports the efforts of local governments and nonprofits to provide pre- and post-natal care, emergency obstetrical services, birth control, and AIDS/HIV programs to desperately poor women living in developing countries around the globe, from Afghanistan to Angola, Ghana to Kenya, Madagascar to Rwanda. It works in countries where an estimated 350 million couples don't have access to family planning; half a million women die annually from pregnancy-related causes and in childbirth; and millions have HIV/AIDS.

The Bush Administration claimed that UNFPA was complicit in forcing women in China to have abortions. In retaliation, it chose to punish all 141 recipient countries by refusing to release $59 million in funding. The administration took that action despite the conclusions of several high-level fact-finding commissions, including its own state department. The commissions found that UNFPA was modeling voluntary family planning in a country desperately in need of that model, and that funding should continue. Cheerleading for the administration was the U.S. Conference of Catholic Bishops, whose president, Wilton D. Gregory, lauded Bush "for taking this action in defense of women and children in developing nations."

So much was at stake that Kissling decided to go to China to see for herself. She gathered together a delegation of nine prominent U.S. ethicists and leaders of religious or faith-based organizations, including the president of Islamic Information Services, the chair of Dartmouth College's religion department, and representatives from the Evangelical-Lutheran, Unitarian, and Episcopal churches, the United Church of Christ, and the National Council of Jewish Women. In six different counties, they visited with UNFPA officials, national and local family planning agency employees, and ordinary Chinese women and men in local villages.

"We believe that the work of UNFPA is critical and . . . highly consistent with our core religious and ethical values," wrote the delegation in a letter to key U.S. senators after their return. "In particular, UNFPA's commitment to voluntary family planning and respect of the rights of individuals and couples exemplifies what is best in our traditions." They added that they had found "no evidence of any coercion toward abortion, and in fact found that abortion rates have dropped significantly" in the counties where UNFPA worked. "We reached the same conclusion as the U.S. State

Department's assessment team," they wrote. "The Bush Administration should fund UNFPA."

Kissling v. the Vatican

Kissling is a globe-trotter, logging hundreds of thousands of miles a year on trips to countries like Mexico and Brazil, the Philippines and Ireland, England, Uruguay, Poland, Italy, Thailand, Russia, Chile, and Puerto Rico, on behalf of the nonprofit organization she leads, Catholics for a Free Choice (CFFC). In her more than twenty years at the helm, Kissling has become a tireless advocate—on television, in print, and at academic, political, legislative, and advocacy gatherings—for the reproductive health and moral authority of women worldwide to make their own reproductive health choices.

Arguably, the Catholic Church, more than any other church in the non-Muslim world, is in a bloody battle to exercise dominion over women's reproductive lives. The Church bases its sexual ethics on natural law, the notion that there is "an unwritten law embedded in all of creation," which humans can decipher through reason and use to build a universally binding moral system. The Church sees natural law as applying to everyone, which is why the targets of its sexual repression are not just Catholic women but all women. This is evidenced by the relentless political activity of Catholic Church operatives in the world to restrict access to birth control and abortion, sterilization and infertility treatments, emergency contraception for rape victims (the morning-after pill) and condoms for AIDS prevention.

It may well be that the Church's most ferocious adversary on these issues—not just in the United States but worldwide—is this barely five-foot, three-inch-tall woman, seated before me on a shimmering red couch.

Kissling has devoted her life to vigorously defending a woman's right to make decisions on reproductive health and sexuality issues. She does this against the ranting of an all-male, authoritarian Church hierarchy struggling mightily to keep women in their place. CFFC runs a host of programs and publishes a magazine called *Conscience*, billed as "a newsjournal of pro-choice Catholic opinion." One function of *Conscience* is to act as a bridge between Catholic women seeking validation for their own moral autonomy and the work of Catholic feminist theologians. Among CFFC's regular contributors is Rosemary Radford Ruether, who has served for years on CFFC's board.

Kissling has been slammed for her boldness. The ultraconservative Catholic League for Religious and Civil Rights calls her "the most notorious anti-Catholic bigot in the nation." She's been labeled a heretic and an enemy of the church. CFFC has been dubbed "one of the most devilish anti-church organizations operating in America" in an American Life League letter to supporters and friends, and twice publicly denounced by the U.S. Conference of Catholic Bishops (USCCB). The USCCB called CFFC "an arm of the abortion lobby" that aims to "promote abortion as a method of population control." It said it merits "no recognition or support as a Catholic organization." In 2002, the U.S. bishops prevailed upon the Democratic National Committee to remove its web link to CFFC's website but the committee refused.

The rage of the Vatican has crashed down on Kissling and CFFC, too. In its nine-hundred-page *Lexicon on Ambiguous and Colloquial Terms About Family Life and Ethical Questions*, released in 2003, the Vatican devoted a full chapter to a vitriolic diatribe against CFFC and its leadership. It denounced CFFC for trying to change the "policies held by the Holy See." It included among CFFC's favored policies "the use of condoms to prevent the spread of AIDS and other diseases," "the use of modern methods of contraception, including emergency contraception, to prevent unwanted pregnancy," and support for "noncoercive, safe, and legal abortion"—which, as a list of issues CFFC cares about, is about right.

In the *Lexicon*, the Vatican accused CFFC of operating a "long-term and persistent program of infiltration and subversion," to which this multibillion-dollar Church with over a billion members worldwide is "extremely vulnerable." It maintained that, since its inception, CFFC "has vigorously attacked and undermined the dogma, teachings and hierarchy of the Roman Catholic Church, seeking to legalize abortion and contraception throughout the world by confusing the faithful. . . . It has caused incalculable damage to Catholics by leading them into sin." It concluded that: "The threat posed by CFFC cannot be overemphasized."

That's a long way from where Kissling began. "They used to say that CFFC was nothing more than one woman with a fax machine," quips Kissling, who is also fond of saying that the Vatican does her best PR work. I watched some of her evolution up close, as a member of CFFC's board from 1991 to 1996. My struggles with my sexuality in the context of Catholicism had driven me years before to write about and work for a recognition of women's moral authority, bodily integrity, and full access to reproductive health care. I'd just finished my first book, *The Choices We Made: 25 Women and Men Speak Out About Abortion*, when Kissling extended the invitation. It was an energizing experience, watching Kissling work. Sometimes I think that if God is a woman, Frances Kissling might be her wrath.

Kissling grew up Catholic in Flushing, New York. Her family was not "a pious family." Her mother was Catholic, her stepfather, German Protestant. They didn't say the rosary or pray together, but she did go to church and Catholic school, developing her own fascination with religion. "My interest in Catholicism was my interest in Catholicism, not my family's interest in Catholicism," she says. "It was the result of my Catholic education, not the result of family formation."

The earliest stirrings of conflict came over Kissling's mother's divorce and remarriage, after which her mother stopped going to church. While Kissling didn't think it mattered to her mother, it did to Kissling. She saw that her mother was not allowed to receive the sacraments, that, "in a subtle sort of way," they were not the "normal family" in the parish. "I know that there is a deep residue of understanding in me, a feeling in me, for the fact that she was treated so badly by the institutional Church," says Kissling.

When she was twelve or thirteen, Kissling began to have conversations about her mother with her confessor. He advised Kissling to have her mother come in to see him. Heartened by the priest's invitation, Kissling encouraged her mother to pay him a visit. She did, mostly out of curiosity, says Kissling. But nothing came of it. "He told her that if she and my stepfather agreed to live as brother and sister for the rest of their lives, she could receive the sacraments. But that she had to receive them in secret, because the parish would still think that she was having sex with my stepfather, and therefore this would be a scandal." Kissling's mother came back and told Kissling that story and was actually "blasé about the whole thing." But Kissling was incensed. She went back to argue some more with the priest, to no avail. Finally she gave up, filing the experience in the back of her mind.

When she began to work with CFFC, those feelings resurfaced. "I am sure that that has a lot to do with the passion I have for what I do. The whole experience of being a child of divorced and remarried Catholic parents, growing up in the forties and fifties, has a lot to do with forming who you are . . . with having a very healthy skepticism about the rules of the Church from childhood," she explains. "Because, look, my mother was divorced and remarried. As far as the Church was concerned, she was an adulteress who was going to hell. As far as I was concerned, she wasn't going to hell, and she wasn't an adulteress. She was a perfectly good person."

After high school, Kissling spent a year in a convent. She fit the classic mode of a working-class Polish-American Catholic girl who went to Catholic schools, she says. Her primary influences were nuns, who were "the best-educated, smartest people I knew." With that grounding, says Kissling, many young Catholic girls "who cared about the world, who wanted to do good and lead a good life" entered the convent. But it didn't take long, a year actually, for Kissling—and the convent superiors—to realize that she did not belong there, and there was a mutual parting of the ways.

Kissling took her skepticism about the Church's power with her into the sixties, the seventies, and the sexual revolution, which were deeply formative experiences for her. "By the time I became a sexually active woman, I had no qualms about myself sexually," she says. "None. Zero. There was not the least bit of suspicion in my mind that there was anything wrong with having sex." Kissling, who never married and works nonstop, became involved in the pro-choice movement and in the operation of abortion clinics. In 1978, she joined the board of CFFC, and in 1982, when the executive director resigned, Kissling moved into that job.

When I interviewed her for *Conscience* on the anniversary of her twenty years at the helm of CFFC, Kissling was surprised to have lasted so long. "I never thought I'd be anywhere for twenty years!" she told me, breaking out in laughter. "I never worked anywhere for more than two years before I came to

CFFC." She referred to herself as a person "who founded things. I started the National Abortion Federation. I opened clinics. I was involved in organizations in their creative phase. . . . I wanted to see if it was possible to stay in an organization and keep the creative drive alive."

She found that it was. At sixty-one, she plans to stay at CFFC. "As long as I'm working, I see no reason not to continue to do this," she says. "I have put in a lot of time, so I have a lot more to offer now than twenty years ago." For the Church hierarchy, that's got to be bad news.

CFFC's Take on Abortion

Between the day in 1974 when one of CFFC's three founders, Patricia Fogarty McQuillan, crowned herself pope on the steps of St. Patrick's Cathedral on the first anniversary of *Roe v. Wade*, the Supreme Court decision legalizing abortion in the United States, and today, CFFC has grown and diversified. What began as a small upstart grassroots organization is now a sophisticated Catholic think tank and the major voice for Catholic pro-choice opinion worldwide. From an entirely voluntary effort to a staff of eighteen, CFFC has been funded by individuals and by major foundations committed to reproductive health, democracy, and human rights, from Ford to the John D. and Catherine T. MacArthur Foundation.

Before CFFC, Kissling explains, it was assumed that all Catholics were opposed to abortion. Since CFFC, particularly under Kissling, that is no longer the case. "One of CFFC's major successes is that we pushed the data about pro-choice Catholics into the public eye," she told me. Kissling and CFFC rallied progressive Catholics to the defense of the country's first female vice presidential candidate, Democrat Geraldine Ferraro, who was denounced publicly by New York's Cardinal John O'Connor for her pro-choice views. Being personally opposed to abortion but committed to upholding the law, Ferraro held the same position as other high-profile male politicians—then New York Governor Mario Cuomo and Massachusetts Senator Edward Kennedy—who were not under such vociferous political assault.

In response, CFFC in 1984 ran a full-page *New York Times* advertisement saying that a diversity of opinions existed within the Church on the question of abortion. "This was the largest public moment on that question in

American history," contends Kissling. "Obviously, very few achievements can be credited to a single organization. They occur in a specific social context, at fertile moments in history, so, in that sense, we're not the only ones responsible. But when CFFC started in 1973, people—especially legislators—believed that most Catholics agreed with the Church and were against abortion. The 1984 *New York Times* ad was a benchmark, a seminal event, in terms of a shift in public recognition that the majority of Catholics were pro-choice."

Indeed, the published petition broke new ground by documenting for the general public the reasons for that diversity of opinion among Catholics about abortion. It included a recognition that no common and constant teaching exists in the Church on what point in a pregnancy ensoulment occurs; a recognition that abortion has not always been treated as murder in canonical history, though it has always been a sexual sin; a reminder of such principles of moral theology as the right to religious liberty (which argues against imposing one's religious views on others); and the Church's belief in the primacy of an informed conscience, defined in Church documents as "the most secret core and sanctuary" of a person, where each of us is "alone with God."

Reprimands by the Vatican were swift and furious. Ninety-seven Catholics had signed that ad, including two priests, two religious brothers, twenty-six nuns, and the rest laypeople, among them prominent theologians. The Vatican demanded that the signers publicly recant or face dismissal from their orders. While the four priests and religious brothers were mildly reprimanded and did recant, the nuns who refused were called ruthlessly and publicly on the carpet. After bitter negotiations, all but two of the nuns "clarified" their positions to the Vatican's satisfaction, noting, for example, that the ad did not support abortion outright. After being raked over the coals by their superiors under pressure from the Vatican, the other two nuns—Sisters Patricia Hussey and Barbara Ferraro—felt they had no choice in good conscience but to leave their communities and vowed religious life.

The Church hierarchy also made examples of signers like theologians Daniel McGuire and Rosemary Radford Ruether, who found speaking engagements canceled or relocated from Catholic venues. Catholic social workers in Los Angeles were ordered by the director of Catholic Charities to stop referring battered women to a shelter operated by Sister Judith Vaughn, who had signed the ad. Accomplished theologian Sister Anne Carr, another signer, was pressured to resign from her official position on the women's advisory committee to the U.S. bishops' committee that was writing the pastoral on women. Mary Ann Sorrentino, then executive director of Planned Parenthood of Rhode Island and among the one thousand signers of a follow-up ad expressing "solidarity" with the original signers, was told by local Church authorities that she had "excommunicated herself."

While the world had come to see that Catholics held varied opinions on abortion, by the beginning of the twenty-first century, an absolutist anti-abortion stand had become the litmus test for loyal Catholics. It had completely eclipsed any such absolutist demand for a Catholic commitment to workers' rights or universal health care, education for all or opposition to the death penalty, the eradication of poverty, or an end to killing in war. All those issues were included, along with opposition to abortion, in the U.S. bishops' statement *Faithful Citizenship: A Catholic Call to Political Responsibility*, published in 2003. But only failing to oppose abortion would qualify a Catholic politician or supporter for public Church sanction and denunciation.

During a bitterly contested battle for U.S. president between the pro-choice, Catholic candidate Democratic Senator John Kerry and the anti-abortion, non-Catholic Republican President George W. Bush, the Church hierarchy escalated its actions against pro-choice Catholic politicians and those who supported them. In 2003, Pope John Paul II issued new guidelines for Catholic politicians, demanding that they defend in their political activity "the basic right to life from conception to natural death," including "the rights of the human embryo." The next spring, Cardinal Francis Arinze, prefect of the Vatican Congregation for Divine Worship and the Discipline of Sacraments, was asked about John Kerry and abortion. He was quoted in the press as saying that unambiguously pro-abortion Catholic politicians were "not fit" to receive Communion, and "if they should not receive, then they should not be given" Communion, either. In June, Cardinal Joseph Ratzinger, prefect for the Congregation for the Doctrine of the Faith, essentially agreed, chastising any Catholic who supports abortion, implying they should stay away from Communion, and authorizing priests to withhold Communion from politicians who support abortion laws and refuse to renege on that position. As for the voter, Ratzinger added in a footnote that voting for pro-choice candidates could be acceptable as long as the candidates had something else to recommend them, that their pro-choice position wasn't the reason for the vote.

The clerical message was reinforced by the efforts of a leading anti-abortion organization. In 2003, on the thirtieth anniversary of *Roe v. Wade*, the American Life League published an ad for "The Deadly Dozen"—pro-choice Catholic members of Congress—in the form of a wanted poster. The next year, the organization listed the names of more than four hundred U.S. pro-choice Catholic politicians on its website, saying they had failed to live up to their responsibilities as Catholics. In 2004, the league took out a full-page *USA Today* ad to chastise American bishops for the "sacrilege" of giving Communion to pro-choice politicians. "You can't be Catholic and pro-abortion!" read the caption below the photo of a priest holding a host and chalice.

Also in 2004, American bishops in Sacramento, California; LaCrosse, Wisconsin; Camden, New Jersey; and Lincoln, Nebraska, either ordered away from the Eucharist or threatened to withhold the Eucharist from pro-choice Catholic political figures—all Democrats. They included former California governor Gray Davis, former New Jersey governor Jim Mc-Greevey, and Senator John Kerry. In the most extreme action, Bishop Michael Sheridan of Colorado Springs banned from Communion any Catholic who would even vote for a politician who supported abortion rights. In response to the public furor that ensued, a special U.S. bishops' task force developed a public statement, which the full body of bishops passed nearly unanimously (183 to 6). It said that Catholic public officials who "consistently . . . support abortion on demand" risk becoming "cooperators in evil" but stopped short of requiring bishops to turn Catholics away at the Communion rail. Instead, it left the decision to individual bishops.

Representative Nancy Pelosi of California—the first woman elected minority leader in the House of Representatives and the highest-ranking woman in the U.S. Congress ever—took a stand against those actions when she spoke at the 2004 March for Women's Lives in Washington, D.C. Before a reported million marchers, she declared herself a mother of five, a grandmother of five, a devout Roman Catholic, *and* pro-choice. Days later, she said publicly that she would not stay away from Communion. Later in 2004, forty-eight Catholic pro-choice members of the U.S. House of Representatives, including Pelosi and three anti-abortion Republican lawmakers, sent a letter to the bishops' task force chairman, Cardinal Theodore E. McCarrick. They warned that denying Communion to politicians who support abortion rights would "revive latent anti-Catholic prejudice" while also bringing "great harm to the Church." McCarrick himself was not in favor of turning pro-choice politicians or Catholics away from Communion. "I'm not going to have a fight with someone holding the sacred body and blood [of Jesus] in my hands," he told the Catholic Press Association.

As for CFFC, it once again became an alternate voice for Catholics nationwide. In a poll of likely voters released in 2004, CFFC found that the vast majority of Catholics did not support the bishops' actions. They said they did not believe that ordinary Catholics or Catholic politicians had to vote on issues in accordance with the bishops' instructions (83 percent), nor did they believe that the bishops should withhold Communion from pro-choice politicians (78 percent), or from ordinary pro-choice Catholics (76 percent). A *Time* magazine poll had similar findings. CFFC surveyed the bishops, too. Of the 145 dioceses either contacted directly or which had issued statements, CFFC reported that the vast majority would not deny anyone Communion or would only do so as a last resort.

Still, to Kissling, this action by some bishops represented a monumental shift. "Those few bishops who have chosen to use Communion as a weapon in America's abortion war have disregarded the long-standing Catholic principles of political freedom and freedom of conscience as they attempt to set forth a new teaching suggesting how one votes on policy measures can be a grave sin," said Kissling in a press statement. "Simultaneously, they are also ignoring the backlash that is likely to occur when they engage in clearly partisan activities."

In fact, nowhere in Church documents are bishops authorized to excommunicate people for being pro-choice. Canon law does say that having an abortion or procuring one for someone qualifies a person for self-excommunication but even then, there are mitigating circumstances. Sara Morello took up the issue when she worked at CFFC. A young Catholic canon lawyer who also spent six years working for the U.S. Conference for Catholic Bishops, Morello made the case that such factors as self-defense, force, or necessity could preclude the punishment of automatic excommunication. She also argued that canon law is not set in stone but has changed, and perhaps most important, that the Church's opposition to abortion has never been declared by any pope to be the highest class of teaching—that is, an infallible teaching.

As for having abortions, Catholic women in 2000 accounted for 27 percent of the women who had abortions in the United States, and they had those procedures at a rate 22 percent higher than the rate for Protestant women. Regarding support for abortion by Catholics, CFFC's 2004 poll found that 61 percent agreed that abortion should be legal, while 38 percent disagreed; among Hispanic Catholics, 62 percent agreed, while 36 percent disagreed.

Polling on the issue is not simple, however, Kissling points out. "Generally speaking—and it depends on how you ask the question—the basic belief among the majority of Catholics is that the decision about whether or not to have an abortion is a decision best made by the woman who is pregnant. The majority has the generic view that abortion should be legal. When you ask follow-up questions, you find that Catholics overwhelmingly support the morality as well as the legality of abortion in situations of rape, risk to the life and health of the mother, and fetal disease. When you get down to abortion for economic reasons, for a teenager, for a woman who just doesn't want to have a child now, Catholics become split almost right down the middle."

The U.S. bishops have helped to demonize women who have abortions, once portraying them in a full-page *Washington Post* ad as having abortions a gurney roll short of the delivery room for reasons like "hates being 'fat,'" "can't afford a baby *and* a new car," and "won't fit into prom dress." Kissling sees women who feel they must end a pregnancy very differently. She believes that for Catholic women, as for all women, the ethical position on abortion begins in a woman's heart. Most women who have abortions "don't buy at all this notion that they're these selfish, atomized individuals who just reject motherhood. . . . Most women still want to have kids. . . . They want their kids to be healthy. They want to be good mothers. And they know that, for the most part, they're having an abortion today because they can't be a good mother today."

She also believes that a certain ambivalence is endemic to abortion. She doesn't think that "being unsettled about abortion is a bad thing. . . . If nobody had any qualms about abortion, I would be worried." Nor is her goal to make everybody comfortable with every abortion. "My goal is that abortion should be legal, and people should be able to make the choice. . . . They should have all the space and all the resources they need to think about this, to wrestle with it. And they should wrestle with it. . . . I don't think the goal of abortion clinics, or abortion counselors, is to say to people, 'There, there, dear. Don't worry about it. There's nothing here.'" To Kissling, the fetus very definitely has value, a value that increases throughout the pregnancy.

As for later abortions, Kissling doesn't worry that too many such procedures are performed but rather that poor women, particularly those whose pregnancies threaten their health or lives or who face severe fetal anomaly, have no access to such care. On the other hand, she says: "If some twenty-two-year-old woman suddenly decided at twenty-four weeks of pregnancy that she'd made a mistake and she just doesn't want to be pregnant and she were denied an abortion, I wouldn't spend five minutes working to get her an abortion. No. I wouldn't care if the law said she couldn't have an abortion. Nobody is fighting for the untrammeled right to have an abortion up to nine months of pregnancy, for any reason whatsoever. This is just bullshit."

CFFC Expands Its Reach

CFFC's reach today extends far beyond the abortion fight in this country. CFFC has confirmed and publicized the extent to which Catholics reject the Church's ban on all forms of so-called "artificial" birth control. In the United States, 96 percent of all Catholic women have used modern contraceptive methods, while only 3 percent use the Church-approved rhythm method, and a majority of married Catholic women have used or approve of modern methods of contraception in many countries around the world, including in Africa, Asia, and Latin America. CFFC has made public the convoluted roots of the Church's ban on contraception, which go all the way back to Augustine. He denounced sex as evil, redeemable only by its procreative potential, and then just barely. To legions of Catholics, the Church lost all credibility on birth control in 1968, when Pope Paul VI rejected the overwhelming recommendation of his own birth control commission in *Humanae Vitae* and maintained the ban. While Paul VI also made a positive contribution by urging the Church to recognize the "unitive" as well as the "procreative" values of married sex, the Church has never reneged on his insistence that every act of sexual intercourse be "open to the transmission of life."

It was Kissling, in the early nineties, who saw the oncoming collision between Catholic health care institutions and women's health advocates over the availability of reproductive health services. *Conscience* magazine ran perhaps the earliest, most painstaking analysis of the issues when "merger mania" took hold, in an article by then CFFC staffer Cynthia Gibson, "The Big Business of Healing." Indeed, when Catholic hospitals began to merge with non-Catholic hospitals, the Catholic facilities demanded that those institutions abide by the U.S. bishops' "Ethical and Religious Directives for Catholic Health Care Institutions." Those directives prohibit many primary care services used by millions of Americans each year. As a result, infertility programs were discontinued. Family planning clinics were jeopardized. Condom education and distribution programs were prohibited. A pregnant woman who wanted to be sterilized right after giving birth, whose doctor had privileges only at a Catholic hospital, would have to make two trips— one to deliver at the Catholic institution, the other to have the sterilization procedure at a non-Catholic facility—at twice the time, twice the cost, and twice the medical risk.

And women who had been raped had to travel not only from the scene of their assault to an emergency room but, if that emergency room was in a Catholic hospital that refused to give them emergency contraction (EC, aka "the morning-after pill," high-dose hormone formulations that prevent pregnancy if taken within seventy-two hours of unprotected intercourse), they had to travel to another emergency room. If they were lucky enough to finally get a prescription, then they might have to travel again, this time in search of a pharmacist willing to fill that prescription. Considered a contraceptive by major medical associations, EC prevents pregnancy by impeding the release of an egg during ovulation, or by stopping fertilization or implantation of the fertilized egg. Because it can prevent implantation, the Church deems it an abortifacient.

Emergency contraception sparked an international furor to which CFFC responded in the 1990s. When the Vatican issued vehement objections to the United Nations providing women brutally raped in the wars in Bosnia and Kosovo with the morning-after pill—who the pope advised should accept the progeny of the rape and make "the enemy . . . flesh of their own flesh"—Kissling's voice was prominent in the roar of opposition that rose around the world. And when the Vatican withheld its annual contribution to UNICEF because of UNICEF's involvement in a health manual that mentioned EC for refugee women who had been raped, CFFC came forward and, in a symbolic gesture, publicly replaced the Vatican's $2,000 contribution.

In the United States, CFFC and other women's health advocacy groups, like Family Planning Advocates of New York State, have brought pressure on legislators to pass laws that would mandate the provision of emergency contraception to rape victims in emergency rooms, in keeping with what has been deemed standard medical practice by the American Medical Association and the American College of Obstetricians and Gynecologists.

The Catholic hierarchy fought mightily against such a regulation, including Cardinal Edward Egan of the New York archdiocese, who made defeating the proposed law one of his main lobbying items on his inaugural trip to the Albany legislature in 2001. But the Church fathers are losing that battle, too. To date, laws mandating the distribution of EC in the ER have been passed in California, New Mexico, Washington, and New York.

In fact, the New York State Catholic Conference in 2003 did a complete about-face and publicly supported the legislation—with the proviso that its hospitals can test to see if the woman is pregnant and withhold EC if she is. In fact, that is nothing but a cosmetic nod to the Catholic directive and a commitment to unnecessary testing. No test will show if a new pregnancy has occurred in the first seventy-two hours after intercourse; it will only indicate a preexisting pregnancy, which would not be harmed by the medication. In Illinois—where information about the medication may be provided but distribution of the actual drug is prohibited—some Catholic hospitals test rape victims to see if they are ovulating and therefore in danger of becoming pregnant. If they are ovulating, these victims will *not* be given the medication or a referral for the medication, even though they are precisely the ones who need it.

Women's health advocacy groups, with the help of CFFC, have organized local communities nationwide to bring pressure to bear on hospitals and governmental agencies in merger situations that threaten basic reproductive health services. In many cases, they have succeeded in saving services as well as in defeating proposed mergers where compromise was impossible.

And CFFC and women's health groups have also been relentless advocates for "contraceptive equity" for women. That means that if a female employee receives prescription drug coverage as part of her health package, that coverage must include contraceptives. In the absence of such equity, women pay as much as 68 percent more in out-of-pocket medical expenses than men do. This, too, has been fought with a vengeance by the Catholic Church, which lobbies for exemptions from such laws based on the concept of institutional "conscience." Again, the Church hierarchy is fast losing ground. By late 2004, some twenty-one states had passed contraceptive equity laws. In California, Catholic Charities of Sacramento challenged the law, lobbying for an exclusion from having to provide contraceptive coverage to its employees, but lost—big. The State Supreme Court voted against them, maintaining that, to be exempt from the law, a religious organization had to have as its mission inculcating religious values, and it had to employ and serve only people who held those values. That is a standard I helped to develop in my published analysis of the impact on women's health of the newer so-called "conscience clauses," which exempt whole institutions from providing what are in effect primary health care services.

It is these actions that have sent some in the Catholic hierarchy to near distraction. At the pitch of the clergy sex abuse scandals in 2002 at the Dallas meeting of the U.S. Conference of Catholic Bishops, Chicago Cardinal Francis George furiously denounced these efforts. Without naming feminists specifically, he referred to "the attack on our health care institutions" and "the attack on our social services through various insurance policies." He also denounced "the opposition . . . to the Holy See's being in the United Nations"—a definite slam on CFFC.

A See Change

On a chilly June morning in 2000, I watched CFFC take its "See Change" Campaign to the East River in Manhattan. Flying the green-and-white "See Change" flag, the schooner *Richard Robbins* sailed at a brisk clip for three hours in the shadow of the United Nations complex. It carried a wind-blown Kissling, her enthusiastic staff, and a cadre of reporters.

Inside the United Nations, the conference called Beijing + Five was under way, where delegates from around the world were reviewing progress toward the goals of the "Platform of Action" created at the Fourth World Conference on Women in 1995 in Beijing.

Those delegates included the Holy See, the government of the Catholic Church, the only religion that has Non-member State Permanent Observer status at the United Nations. That means that unlike representatives of any other religion, the Holy See has the privilege to speak and vote at UN conferences. The goal of CFFC's cleverly named (by Kissling) "See Change" Campaign is to change the Catholic Church's status at the United Nations from a government to what all the other world religions are: a nongovernmental organization. The spectacle, the passion behind it, and the worldwide receptivity to the See Change Campaign are arguably what put the Vatican over the top on the subject of CFFC.

CFFC's See Change Campaign had its roots in the increasing frustration among women's health advocates over the highly contentious and obstructionist stance taken by the Holy See at the world's most important meetings on women's rights, beginning with the UN's International Conference on Population and Development in Cairo in 1994. "Why should an entity that is in essence one hundred square acres of office space and tourist attractions in the middle of Rome, with a citizenry that excludes women and children, have a place at the table where governments set policies affecting the very survival of women and children?" said Kissling to *Ms.* magazine. "Nobody else at the UN claims to represent God."

Kissling has personally witnessed the ways in which the Vatican has stalled consensus on issues vital to women's reproductive health, joining forces against women's health advocates with some of the most reactionary and fundamentalist regimes in the world. She has witnessed the Vatican fighting to exclude any mention of the word *contraception* from UN documents. She has watched the Vatican fight against affirming the rights of individuals and couples to family planning, and denounce the promotion of condoms for AIDS prevention. She has seen its representatives reject the notion of reproductive rights as human rights because that could mean approving abortion and sex outside of marriage. She has witnessed their trying to remove "forced pregnancy" from a list of war crimes because it might be used to support abortion. And she has seen them rail against the demand that where abortion is legal, it should be safe. "You actually physically sit there and watch them go over to the delegation of Libya and make common cause with the most regressive, totalitarian regimes in the world, for political expedience on the issue of abortion," reports Kissling.

Despite the Holy See's vigorous opposition to reproductive rights issues and its blatant hostility to the involvement of "Western feminists," the UN meetings revolutionized the way women's reproductive rights are seen in the international arena. The Beijing Platform of Action reads in part: "The human rights of women include their right to have control over and decide freely and responsibly on matters related to their sexuality, including sexual and reproductive health, free of coercion, discrimination, and violence."

"When one reviews the history of state and religious attempts to control women's sexuality," Kissling has written, "such a statement in a governmental document is nothing short of revolutionary."

Despite its critics—including the U.S. House of Representatives, led by avid conservative Catholic pro-lifer Representative Chris Smith (R-NJ), which passed a resolution denouncing the campaign that, to Kissling's relief, never made it out of the Senate committee—the See Change Campaign has been a great symbolic success. In addition to an avalanche of worldwide press, Kissling reports that tens of thousands of individuals, about eight hundred groups, and representatives of several countries have endorsed it. "Politicians from the United Kingdom, Sweden, Ireland, and Germany—a total of fifty European parliamentarians—signed a letter that went to all the UN ambassadors telling them to pay attention to this campaign," says Kissling. "And we were invited by members of the European Parliament to organize with them a seminar in the European Parliament on the role of religion in public policy, with specific information on the See Change Campaign." At that seminar, Kissling reports, "We had over one hundred people attend."

Going Global

CFFC's international mandate had its roots in the organization's discovery over time that people in other parts of the world, particularly Latin America, knew what CFFC was and had begun to use their publications. There came a recognition, explains Kissling, "that CFFC was Catholic first and American second. The same way the Church is universal in its teachings and its reach, the ideas and the pro-choice perspectives of CFFC have resonance with Catholics all over the world." Today, CFFC has partner organizations in seven Latin American countries (Argentina, Bolivia, Brazil, Chile, Columbia, Mexico, and Peru); and works as well with colleagues in Canada; Europe (Ireland, Poland); and Africa (Kenya, Nigeria, Zimbabwe, and Uganda).

This movement into the international policy arena has increased CFFC's power exponentially. It has also required CFFC to see its issues through a wider lens, as Kissling explains it: "that it's not just about abortion; it's about women's rights. It's about women's rights, but it's also about sexuality. It's not just about sexuality; it's also about institutional power. And today, it's not just about Catholicism; it's about religion." That is, fundamentalism, a major target of CFFC's work in recent years. "It is not just a question of the role that Catholicism plays in the world; it's the role that Islam plays, that Conservative Judaism plays. It's the ways in which governments ... use the conservative elements of religion as mechanisms of control," continues Kissling. She sees the dramatic rise in conservatism and conservative religious institutions since 1980 as inspiring "an enormous backlash worldwide against liberalism, the women's movement, and modernism. That push to conservatism, to fundamentalism, includes the control of women. Always."

One of CFFC's most provocative international initiatives began on World AIDS Day 2001, when CFFC hoisted billboards, mounted posters, and ran print ads that turned the Church's commitment to a "culture of life" on its head. Wearing tall miters and white chasubles over red robes, six somber Catholic bishops marched forward on one poster, clutching their prayer books. Above them read the words: "Catholic people care." Across the center: "Do the bishops?" At the bottom, in white letters on a red background: "Banning Condoms Kills."

CFFC aimed its Condoms4Life Campaign at "defending the sanctity of life." The precipitating event for this campaign was the refusal of South Africa's bishops to change their public position on condom use. Indeed, in sub-Saharan Africa, 25 million people are infected with HIV, out of 37.8 million worldwide. In South Africa, some 5.1 million people are living with AIDS, and the overall HIV infection rate is nearly 22 percent. In 2003, AIDS killed 2.2 million Africans alone, and left 12 million children orphaned. And, as UN Secretary General Kofi Annan has attested, "AIDS increasingly has a woman's face." Women make up 50 percent of those infected with HIV worldwide, and 57 percent in Africa. In the United States, women represent 30 percent of all new HIV infections. Women of color suffer the most drastically; they account for fully 82 percent of all new infections among women (64 percent black women and 18 percent Hispanic).

The appeal for change in South Africa came not from the Church's "enemies" list but from a devoted insider, Bishop Kevin Dowling of Rustenburg, South Africa, coordinator of the South African Bishops' AIDS Office. Dowling issued a statement before the bishops' 2001 meeting arguing for a change in position on condoms. He acknowledged his own "anguish over the enormity of the suffering of people in the AIDS pandemic ... which I have experienced in a very personal way in my own ministry." Dowling argued for looking at condom use "not as a means to prevent the 'transmission of life,' but as a way to prevent the 'transmission of death.'"

Despite the devastating human toll of the disease in Africa and other parts of the developing world, in 2001, the South African bishops confirmed traditional Church teachings. In a statement paradoxically entitled "A Message of Hope," they declared that "widespread and indiscriminate promotion of condoms [is] an immoral and misguided weapon in our battle against HIV/AIDS. . . . The use of condoms goes against human dignity. Condoms change the beautiful act of love into a selfish search for pleasure, while rejecting responsibility. . . . Condoms may even be one of the main reasons for the spread of HIV/AIDS."

Other powerful members of the Catholic hierarchy have viciously denounced condom education and use. In a flabbergasting misrepresentation of the facts, Cardinal Alfonso Lopez Trujillo, the president of the Vatican Pontifical Council for the Family, told the BBC in 2003 that "the spermatozoon can easily pass through the 'net' that is formed by the condom." He chastised health ministers to handle condoms the way they handle cigarettes, by warning that they are a "danger" to the public health. The international community responded with outrage and alarm. Peter Plot, head of the Joint UN Programme on HIV/AIDS (UNAIDS), said: "When priests preach against using contraception, they are committing a serious mistake [that] is costing human lives. We do not ask the Church to promote contraception but merely to stop banning its use." The World Health Organization denounced the Vatican's views, too. "These incorrect statements about condoms and HIV are dangerous when we are facing a global pandemic, which has already killed more than twenty million people," an official told the BBC. A special panel of the National Institutes of Health concluded that condoms are "essentially impermeable" to even the smallest of STD viruses. And UNAIDS reports that consistent and correct condom use reduces the risk of HIV infection by at least 90 percent.

Despite the reactionary decision by the Southern African bishops, Dowling remains a passionate advocate for a complete rethinking of Catholic theology around condom use and AIDS. Dowling maintains that wearing a condom to prevent the spread of AIDS is not a contraceptive action but a lifesaving action, not an "intrinsic evil," as the Church has labeled all forms of artificial birth control but, theologically speaking, "a moral imperative."

In vocal support of Dowling were more than sixty nuns calling themselves Sisters for Justice of Johannesburg. They charged that, for women particularly, there is no hope in the Southern African Catholic bishops' "Message of Hope." Calling the bishops on their naive notions about sex, described by the bishops as "the beautiful act of love" between "equal and loving partners," the nuns insisted that the bishops' message was directed to people in "fairly healthy and stable marriage relationships." It failed to address "people, usually women, in abusive, oppressive, or desperate relationships or circumstances and who are very much at risk of being infected by the HIV."

They noted that "the greatest incidence of new HIV infections is in young women in the fourteen to nineteen-year-old category, primarily due to the high incidence of forced or reluctant sexual intercourse." They said, "It is imperative that women and girls be provided with the information and the means to protect themselves." They quoted with enthusiasm Mozambique's prime minister, who argued at the 2001 UN Special Assembly on AIDS for further development of HIV prevention methods like female microbicides (tasteless, odorless, intravaginal products) that women can use "with or without a partner's knowledge or cooperation." The nuns also advocated greater use of female condoms.

The bishops did make one change. They allowed married couples where one partner is infected to use a condom. The nuns saw the contradiction there, insisting that everyone has "the right to protect their life against mortal danger," not just the uninfected married partner. Furthermore, they implied, if the bishops maintain that condoms don't work, then why are they allowing anyone to use them to prevent the spread of a disease they don't think it will contain?

For World AIDS Day 2002, CFFC's Condoms4Life Campaign commended Bishop Dowling and printed up hundreds of postcards that supporters of the campaign could mail to him, which read: "Thank you, Bishop Dowling, for showing compassion, leadership, courage, and good sense in the fight against HIV/AIDS." They also lauded the Sisters for Justice in Johannesburg for the principled and unequivocal stand they took in Dowling's support. And they commended the bishops' conferences in France and Germany, which claimed condom use could be a moral good in some circumstances.

In 2003, CFFC expanded the theme of its Condoms4Life Campaign by urging "sexually active Catholics, especially young people, to use condoms as part of a mature, responsible sexuality." In direct defiance of the Church's official position, the ads have a banner headline that reads: "Good Catholics Use Condoms." CFFC's main message is best illustrated by a poster that features the smiling faces of four embracing couples—white, African-American, Hispanic, straight and gay—and says: "We believe in God. We believe that sex is sacred. We believe in caring for each other. We believe in using condoms."

In addition to mounting posters in Washington, D.C. (where the Catholic Church failed to generate enough pressure on the Metro Transportation Authority to bring them down), Condoms4Life materials have traveled to Belgium, South Africa, Kenya, Zimbabwe, Chile, Mexico, and the Philippines. CFFC's goal is "to move the campaign around the world . . . letting Catholics know that there is a Catholic perspective on condoms and HIV other than that being offered by the Catholic hierarchy."

In 2003, the ultraconservative Catholic League resurrected and publicized an old Kissling quote, evidence of how much combing they're willing to do to attack her. It said: "I spent twenty years looking for a government that I could overthrow without being thrown in jail. I finally found one in the Catholic Church."

That statement makes me uncomfortable. I assume it might be something Kissling is sorry she said. I decide to bring it up when I interview her for this book. When I do, I make the mistake of calling it "that awful quote."

Kissling strikes back. "I think it's a great quote!" she says, shifting her weight and raising her voice. "I think it's an absolutely great quote!" She says that it "very much signifies who I am." It speaks to the fact that "I spent my entire life as an advocate for social justice." She has "every pride" in that quote "whatsoever."

Digging myself in deeper, I say: But the way it reads, it sounds like this arbitrary person who's out there wanting to topple regimes.

"No, it doesn't! It doesn't at all!" insists Kissling. "First of all, it talks about the Roman Catholic Church as a political institution. It's a government. I didn't say I want to bring down the Roman Catholic Church. I said I was looking for a government that I could overthrow that couldn't put me in jail. Okay?"

She's not even thinking of stopping. "What do those few words encapsulate? They encapsulate . . . the powerlessness of the Roman Catholic Church," she says, "a government that can't put you in jail. . . . It's not really even a government. What can they do to you? Nothing." To Kissling's ear, that quote says: "I found in the Roman Catholic Church an unjust political system that I can and am committed to overturning. And that system is actually very weak."

Then Kissling draws an analogy, even though she admits it is a dangerous thing to do. I think she is comparing the response to the provocative work of CFFC from both inside and outside of the Church reform movement to the response to Jesus' behavior in his own time. "This would be like people screaming and yelling that Jesus Christ was too much of a radical because he threw the money-changers out of the temple," she says. "He didn't do it in a very polite way. He went in with a whip. He turned over their tables. . . . He ranted, he raved. He screamed, he yelled. You know, I think the Catholic League would be very unhappy with Jesus Christ. . . . Jesus Christ wasn't sitting around trying to figure out which issues were safe, which issues wouldn't get you in trouble."

"What is social change about?" she asks rhetorically. "Remember, it's a reform movement. Within the [Church] reform movement, you have different strategies, just as you have different strategies within the women's movement. It's why we have the National Women's Political Caucus, Emily's List, the Fund for the Feminist Majority, NOW." She says that CFFC's "approach to social change is prophetic. The prophetic tradition is a tradition of provoking. There is very little in the reform movement that is prophetic."

When she's done, I feel like I've been hit by a train. This side of Frances Kissling, even for a few minutes, is not a great place to be. Which is why some people have a hard time dealing with her—foes and some friends, too. Theologian Ada María Isasi-Díaz calls CFFC "a pain in the rear of the Church," but upon closer scrutiny, she's actually talking about Kissling. "How can Frances not be a pain in the rear?" she says, laughing. "She's a pain in the rear to everyone. I love her dearly [but] she's like the pope. Frances is the pope of the Left. She's very much a centralist, too. . . . She's very much in control of this organization. It's going to be what she says. But I think she has done wonders in making the issue be out there, be alive, extended it. And in a certain way, also, keeping the focus."

In 2004, Kissling broke with the pro-choice movement and made an impassioned argument for reframing the pro-choice position to take into account not only the value of a woman's fundamental right to choose but also the competing value of a developing human life. In a seminal essay in *Conscience*, she acknowledged the well-grounded fear among pro-choice advocates that, if they accord the fetus value, it will lead inexorably to demands that all or most abortions be prohibited, but she said that it was time for the pro-choice movement to take that risk. "The pro-choice movement will be far more trusted if it openly acknowledges that the abortion decision involves weighing multiple values," she wrote, "and that one of those values is fetal life." Kissling went a step further, advocating a move from acknowledging that the fetus has value to examining the obligations for pro-choice people "that flow from that value"—such as dealing humanely rather than defensively with the matter of fetal pain as it pertains to abortion practice.

Whether that position will help close the rift between pro-life and pro-choice forces—including in the Church reform movement—remains to be seen. For while progressive reformers may support birth control and fertility treatments and condoms to prevent AIDS, many part company on abortion. That makes the subject a cause of great strife in the ranks and a continuing challenge to Church reform.

7

STRIFE IN THE RANKS: THE ABORTION CHALLENGE TO CHURCH REFORM

Sister Joan Chittister is opposed to abortion. But she presents her own challenge to the Church's position. "I'm among all the women in the world who would say I do not believe in abortion as a birth control method of choice," she told me. But "I will never, ever condemn a woman who has had an abortion. Why? Because of what the Church has taught me."

She explains that "the Church has nuanced every single life situation. The Church says you can kill for the state; you can kill in self-defense; you can kill to punish. There isn't one life question that the Church has not nuanced when life is in the hands of men. It can always be modified, nuanced, justified." And that, she points out, is not about "one life at a time," but about "all life on earth! . . . They say nuclear weapons are a matter of theological doubt. There might be some circumstances in which they would be acceptable." By contrast, she goes on, "When life is in the hands of a woman, it can never, ever, under any circumstances, for any reason whatsoever, be contravened. . . . I say you can't have it both ways." To her mind, "our life ethic here, our morals here, are inconsistent."

Chittister's recommendation is to put that question "to the assembly of the faithful to discern." She asks herself rhetorically: "Am I opposed to abortion? I am. Am I opposed to the question? I'm not. I want you to make it a matter of all life questions. Don't just hang women out there, and hang all the life issues on women's shoulders."

Moral theologian Christine E. Gudorf favors legal abortion but holds a similar view. She went a step further than Chittister to illustrate the different approaches the U.S. bishops take to the various "life" questions. She took the bishops' pastoral letter on nuclear war and turned it into a pastoral on unwanted pregnancy, maintaining the original tone and spirit. Unlike abortion, where the moral issues are black, white, and absolute, she wrote that in the bishops' pastoral, "We find no judgment on persons, no condemnation of those who make the wrong choice, no list of specific circumstances under which a pilot can decide to bomb a target or a soldier to shoot." She maintained that the bishops wrote the pastoral letter "to provoke reflection and discussion," based on "the conviction that only such reflection and discussion will lead to the end of war." Gudorf asked, "Why is persuasion deemed appropriate for dealing with one kind of social policy regarding the taking of innocent life but inappropriate for another?" The way it looked to her, she wrote: "Women with the medical option of abortion seem, in the bishops' eyes, to be a greater danger to life than men armed with tanks, missiles, and bombs."

Gudorf also wrote of adopting a child who had multiple birth defects, and later conceiving a child in a pregnancy that threatened her own health. During that pregnancy, she developed rheumatic fever, which led her doctor to advise immediate abortion. But Gudorf chose instead to carry the pregnancy to term. Yet, because she was against making abortion illegal, she was described by anti-abortion advocates as being "pro-abortion." "I am not pro-abortion," she wrote, characterizing that accusation as an oversimplistic, inaccurate statement of her advocacy for a woman's right to choose.

Speaking Across the Divide

There is no question but that abortion remains a contentious subject within the progressive Church reform community. Some of the women I spoke to define themselves as pro-life; others reject a legal ban on abortion but have grave difficulty with the issue; still others won't talk about abortion at all, for fear of Church reprisals.

Maria Coffey is a perfect example of just how complex abortion is in Church reform circles. At sixty, she's retired. She had been dreaming of becoming a volunteer Eucharistic minister, quietly serving the Church she loves. But when the clergy child sex abuse scandal broke in 2002, she was so devastated that she became what she had never been before: a Church activist. Coffey is a cofounder and former coordinator of the West Side Voice of the Faithful chapter in Manhattan. When I visit Coffey, she is on the planning committee for the Northeast Regional Conference of Voice of the Faithful, which will turn out to be a rousing success, drawing some 1,500 people. She welcomes me warmly into her son's former room in her vintage Manhattan apartment, offers me a mango fruit drink, then declares: "I warn you, I'm conservative."

Coffey may be conservative, but she describes herself as against mandatory celibacy, which she thinks is implicated in the child sex abuse crisis. She is appalled that only "a little organ" stands between Catholic women and the priesthood. She is anticlerical, eager to dispense with all those "princes of the church," to bring back the peripatetic priests of old who knew and cared about their people, and who used to gather around dining room tables on Sunday afternoons in homes like the Coffey's, in 1950s Boston.

A divorced and remarried Catholic, Coffey is adamantly opposed to the Church's annulment process, which she had to endure personally before marrying her second husband. On the subject of homosexuals, she admits she came to acceptance late, in part due to her sexual naïveté about what homosexuality actually was. Today, she says, "Who are we to judge anybody? I would hope no one would judge me because I got a divorce, and I would not judge anybody because of their sexual orientation."

She completely supports the use of birth control because it does not interfere with the development of a life and believes that Church authorities "should stay out of that." She thinks "minutiae has crept in." She says she knows that the Church hierarchy is "trying to set guidelines, but it's a bunch of old men, or a man . . . getting hung up on little rules and regulations." Besides, she adds, "I can't see Christ sitting there going, 'Let me see. Is an IUD better than a diaphragm?'" She'd like the hierarchy to have more faith in people. "It seems to me that what we need are real, essential guidelines for how to live a good, productive, Christian, soulful, spiritual life. And then, maybe trust us to kind of fill in the details."

But Coffey is completely opposed to abortion. "To me," she says, "it's murder." And, she adds, "If you think it's murder, it's [murder] under any circumstances." She has not come to this position without careful self-examination. "Believe me," she says, "I've gone through all kinds of stuff with this." What she's come to is that "since we don't know whether a soul in the person begins at the moment of conception . . . we have to hitch our bets on the side of, if it's a soul and a person, you protect it. . . . If God were to send down a new commandment that said, 'You know what? You're wrong about this. Humans don't start until . . .' then, fine. [But until then] I think you hedge your bets on the side of life." Indeed, one of the major arguments for differing views of abortion among Catholics is the theologically outstanding question of when ensoulment occurs. Historically, there was agreement that it did not occur at the moment of conception (now recognized as a process that takes place over time, not a single moment), but at a later point in the pregnancy.

"I also believe," she added, "and this is where I am going to sound conservative, that the choice . . . begins when that man and that woman decide to have sexual intercourse that's unprotected. I don't believe choice should be at the point where you're deciding whether or not to abort a possible child. That sounds judgmental, and I know there are circumstances, certainly rape and certainly incest, and other situations where that comment is totally erroneous and inapplicable. But I think there are many times . . . when the choice should be made at that point, not at the other point."

So is she willing to allow abortions in those cases? "If I believe that it's a soul and a life, I just don't know where you make exceptions," she says. "It's a tragedy. It's a terrible, terrible tragedy." I start to ask: "Do you punish a woman by forcing her to carry an unwanted pregnancy to term?" But I only get as far as "punish" when Coffey snaps back: "Does it have to be punishment to the person bearing the child?" "But what if it's a child of rape?" I ask. "You can give it up for adoption," she says.

As I ponder the clarity of Coffey's position, I hear her adding, rather ruefully and to my surprise, that she knows "all the other sides of this." As it turns out, a dear friend of Coffey's had an unwanted pregnancy. The friend chose to have an abortion and turned to Coffey for aid. "I helped her, once she had made the decision," explains Coffey. "I did not agree with her decision, and she knew that." So, confesses Coffey, "I'm not pure on this. I didn't have to help her." Illustrating the ambivalence that can intrude on even the most deeply held beliefs about abortion, Coffey adds: "There are so many gray areas. I don't know. I feel that maybe I wasn't true to [my beliefs about abortion], but I was true to my other part of myself, which is that this friendship meant a lot, and she needed help, and she had made her decision. We all make mistakes. So I made my choice."

Now I really want to know more. Since Coffey rejects the Church's taking a position against birth control, I wonder if she thinks it's okay for the Church to take such an absolute position against abortion. I ask. Suddenly, Coffey turns the tables on me. "This is tough for you, isn't it?" she asks.

At first, I feel defensive. Then I acknowledge, it is. I know about those gray areas, too, but from the other side. I believe in a woman's moral right to terminate a pregnancy, but I also believe that the fetus has value. I find pictures of the developing fetus, in relationship to abortion, troubling. I care about fetal pain and believe that abortion practice must address that. I want all abortions done at the earliest possible moment, in the first trimester of pregnancy, and I want there to be as few as possible. To achieve that, women must have everything they need to prevent a pregnancy by way of birth control and sexual safety. If contraception or common sense fail, women need immediate access to over-the-counter emergency contraception, the availability of which the Catholic Church and the Bush Administration are fighting tooth and nail. If it is too late for that, women need unfettered access to the earliest possible medical or surgical abortion procedures. But fundamentally, I believe that making abortion illegal will not make it go away; it will just kill more women. It stuns me that 68,000 of the world's poorest, most disenfranchised women die from unsafe abortions each year—more than the number of American soldiers killed in the entire Vietnam War. I envision the women's bodies lined up, row upon row.

"I think that the big picture has been laid out by Christ, and it's his infinite love of human life," says Coffey, summing up her position. If you take that as your starting point, you will conclude that "life is precious. Life is godly. Life is a soul. And life has a right to be," she explains. "If you believe that," she continues, "then you can't be for abortion. You can't. It just doesn't make sense. It's like, who am I as a mother in any kind of distress, whatever the distressful situation, who am I to value my life over the life of this child that may end up saving the world?" It is a provocative question. I respect her answer.

Eschewing the Labels

Confounding extremists on both sides, many Catholic women, especially young women, consider themselves pro-choice *and* pro-life—an important position for which we have yet to give a label. That is how a young teacher, Michele Curay-Cramer, described herself in press accounts of her firing from a Catholic school for signing a pro-choice petition on the thirtieth anniversary of *Roe v. Wade*, in 2003. "You can believe that a woman has a right to choose, and you can also believe that abortion should be the last possible option," she told the local press. This is also how many young women responded to a poll I oversaw when I briefly directed the Pro-Choice Public Education Project, back in 1998.

Jenny, a progressive Church reformer and youth minister at a Catholic parish who fears reprisals if she uses her real name, feels similarly. "Abortion is a really hard one," she tells me, "especially this week. We have a sixteen-year-old girl at our parish. She just found out that she's pregnant and is definitely leaning toward abortion. I can't tell her what to do. . . . But that's hard to hear."

Jenny says that, back in high school, she was "a big pro-lifer," but she feels differently today. She sees abortion as "a hard thing" because she believes that it is the woman's "private choice," but she also believes that "the fetus is a human being from conception." Of the young pregnant woman, Jenny acknowledges that "it can be an embarrassing thing, to be pregnant and be six-teen." But Jenny also sees adoption as "such an amazing option." In her mid-twenties, Jenny knows plenty of people waiting to adopt.

Jenny and the other youth minister decide to support the young woman, "in whatever she does." While Jenny says "absolutely" she could jeopardize her job if she took that stand publicly, in this case, "the discussion was in a counseling setting," and she feels that "pastorally, my response was appropri-ate." Jenny also finds comfort knowing that there are resources to help the teenager deal with the aftermath of the abortion—something Jenny believes the young woman will have to do, sooner or later. As to what will become of the young woman spiritually, Jenny is clear. "I don't believe at all that if somebody has an abortion that you've sinned forever, and you're never going to be able to get to heaven. I don't believe in hell, period. That's not an option that I think God has for us. I think God is more loving than that."

Kate Ott is a twenty-seven-year-old, married, Catholic mother of one, who is also searching for that middle ground. A Christian ethics doctoral student at Union Theological Seminary, she is also a member of Persist, Union's feminist caucus, which lists as its agenda "to persist in resistance to discrimination against women," and to celebrate "our foremothers' lives and theological vision."

On a cool March night in 2004, the members of Persist sponsor Cyn-thia Cooper's pro-choice play, *Words of Choice* (which includes two excerpts from my book *The Choices We Made: 25 Women and Men Speak Out About Abortion*). Their orange flyer says that Persist "has taken no stance on abor-tion," expects disagreement, and hopes that the disagreement can be ap-proached in a spirit of trust, love, and justice. It also says, "Being a feminist should be about passionately caring for whole people rather than becoming ideologues."

Ott is on the panel that follows the show. Of abortion, she says, "Any ethically sound choice will never be easy. It won't make you feel really good, and it won't make everything all better." She thinks that the polarization into two distinct camps is the heart of the problem. She sees a great need "to tell stories," but they "have to be stories that confound" the categories. One can only do that, she explains, if people will agree that "these are not actual spaces that exist," that the camps in reality are not so mutually exclusive, and that within one you will find components of the other.

She takes as an example the most contentious public policy issue: late abortions. She talks about a woman who faced the terrible choice of continuing or ending a pregnancy of a wanted child with grave malformations. The woman's physicians told her the chances her child would die were 99 percent. The issue for the woman, posed by Ott, was not whether her child would die but "when and how." She thinks that circumstance is analogous to family members who, faced with a terminally ill relative, "have the right to shut off life support." She doesn't think there is a difference between those two situations. Furthermore, she insists, "It is the stories like that that pull you into the middle. They are the ones that need to be told the most."

Claire Noonan, a thirty-two-year-old Catholic chaplain at Loyola University in Chicago and former spokeswoman for national Call to Action, has seen the warring factions line up on campus. She reports that there are two groups of students. One consists of those who are theologically and politically conservative, evangelical, and members of the Life Teen Movement, Youth for Christ, and the pro-life movement. The other group is made up of the "students against sweatshops," those involved in faith-based actions and antiwar activities. Noonan struggles "to bring them together, to break down the stereotypes they have of each other, that the justice kids don't care about their faith, and the conservatives don't care about people." Those stereotypes, she says, are "not true."

Indeed, Rea Howarth, national coordinator of Catholics Speak Out—a program of the Quixote Center focused on "equality, justice . . . and dialogue between the laity and the hierarchy on issues of sexuality, sexual orientation, and reproduction"—sees a convergence on the horizon, too. She has been a pro-choice advocate for years. She thinks the day for real dialogue may be at hand. "Catholics do not want to see abortion ended. They want that legal option there," she says. "But they're deeply conflicted about it and would like to see some restrictions, post first-trimester." Howarth doesn't think limiting abortion to medical necessity after the first trimester is such a bad idea, though she tells me she is sure that saying that publicly "will get me into hot water with some folks. But I really don't care. There are legitimate concerns about having abortions after the first trimester." At the same time, she says, "If somebody tells me, 'You're committing infanticide if you have an abortion at six weeks into a pregnancy,' that's bullshit." Between those two poles, Howarth thinks "there ought to be some middle way, some middle ground, where as a society, we can have a discussion about it."

Avoiding the "A" Word

At the same time, some major Catholic social justice organizations completely avoid the contentious reproductive health issues. That has been the case since the founding of Network, the highly regarded, women-run Catholic social justice lobby founded by Catholic sisters in 1972. Dominican nun Maureen Fenlon, who took over as national coordinator in 2003, attests that Network is pressured from pro-life forces, including the Church hierarchy, as well as pro-choice forces, to take a side on the "life" issues but consistently refuses. Network's focus is economic justice, issues from a living wage to a strong social security system, affordable housing to just treatment for immigrants to fair taxation. Says Fenlon, no matter what people's views are on the abortion issue, and it ranges within Network, "it was not a desire ever to get embroiled" in it.

Network has chosen instead to be a "voice for the poor" on Capitol Hill, particularly urgent at this time, says Fenlon, when there is no other voice. Indeed, she faults the institutional Catholic Church for failing to call Catholics to action on these critical issues.

"Catholic social teaching is the Church's best-kept secret," says Fenlon, referring to the body of Church teaching based on papal encyclicals and bishops' statements in defense of the poor, workers' rights, economic equality, environmental justice, disarmament, and peace. Unfortunately, says Fenlon, most Catholics haven't a clue about these teachings because they're not being touted and preached from the pulpit. In reality, she admits, people don't necessarily want to hear about these issues, which challenge "the social frameworks"—economic, political, medical—under which we live. "If we were to have a Church that spoke in these terms, they would be challenging their very friends who run the banks, run the corporations, run the show."

But even the broadest-base Church reform group, Call to Action, has not in recent years had abortion on the agenda of its highly attended national conference. At the end of its 2003 conference, two young women—one a volunteer escort at her local Planned Parenthood and the other a youth minister at her Catholic college—sat disheartened that there had been no place to go to discuss not just abortion but the whole area of sex and women's reproductive health. In 2004, though Call to Action chose as its theme "Sex, Science, and the Sacred: Embracing Divine Mystery," not a single workshop had as its subject abortion.

Because it has a narrow three-part agenda—supporting victims of abuse, priests of integrity, and structural change in the Church—Voice of the Faithful would not be expected to deal directly with abortion. But even that organization has had to wrestle with the issue. After its first national conference in Boston in 2002, accusations of being a supporter of abortion led to VOTF's public renunciation of Debra Hafner, a highly regarded ordained Unitarian minister with twenty-eight years of experience as a sexuality educator who had led a well received breakout session on how to keep children safe in faith communities. In response to complaints that this former head of the Sexuality Information and Education Council of the United States and onetime educator with a Planned Parenthood affiliate supported abortion, VOTF's president posted a letter to the organization's then twenty thousand members apologizing for having had her speak—an action to which Hafner responded with great dismay. At VOTF's 2003 New York City regional conference, the subject of abortion came up in the youth panel, which featured five young Catholic students. From the audience and, ironically, from a self-described pro-life panelist came impassioned concerns that many young, pro-choice women are leaving the Church over its unequivocal opposition to abortion—evidence of the longing from all sides of the aisle for some kind of consensus.

In fact, the U.S. bishops may be inadvertently supporting an even stronger willingness among Catholics—including the vast majority of bishops who have not taken the harsh stand of turning pro-choice Catholics away from Communion—to respect difference and reach a consensus, however uncomfortable, on abortion. While the debate rages, the statistics show that in fact Americans, including Catholics, have begun to settle, clearly with unease on both sides, into a compromised middle ground where the majority favor abortion at least in some circumstances and reject making the procedure illegal. By polarizing the debate even further, using abortion as the litmus test of Catholicity and politicizing the Eucharist, the U.S. bishops may be enlarging that middle ground.

But Why Abortion?

While the issue itself is compelling and controversial, the prominence with which it is treated by the Church, particularly in 2003 and 2004, raises an interesting question: Why is the greatest sin of the early twenty-first century in the Catholic Church not nuclear proliferation or ethnic cleansing, prisoner torture or blowing children up with bombs or land mines, hungry people in the richest nation on earth, homelessness, rape, murder, or priest abuse of children, but ending a pregnancy, even at the zygote stage? It is true that the Church hierarchy and the Republican Party have developed a much closer alliance than they have ever had before in the United States, and one can conjecture that preserving that alliance between the Church and like-minded politicians on abortion is a factor. But I think there are two other factors that are equally important.

One is, simply, that the women who have abortions are far easier targets than, say, the men who wage war or build nuclear arsenals. The bishops would have a much harder time demonizing those men and exercising absolute moral judgments over them; such a position would be seen as overly simplistic, misguided, and foolish. In fact, the abortion debate is a profoundly patriarchal one. The flesh-and-blood Catholic women who have abortions are all but invisible. In their place are women as victims—misguided by big bad legislators or crazed scalpel-wielding doctors; ditzes, who have abortions because otherwise they "won't fit into prom dress"; and demons. Yet it is Catholic women—not their legislators or their doctors—who make their own moral choices every day to end pregnancies they cannot handle, and then sit in the pews beside everybody else. The hierarchy doesn't even give these ordinary women the courtesy of wrestling with them directly. For this to be a real moral argument, the bishops would have to recognize, address, and debate those women. That can't happen, of course, because women—including pregnant women—have no place at the hierarchy's celibate, all-male table, where the ultimate decisions about the interpretation of Church teachings on women, sex, and childbearing are made.

In addition, I think that targeting abortion is part of a larger effort by the Church hierarchy to regain its moral authority by renewing its power over the Catholic family. But the Church cannot exercise authority over the Catholic family without the cooperation of women. In its 2004 document "On the Collaboration of Men and Women in the Church and in the World," the Vatican chastised women who dared to seek equality through "liberation from biological determinism." While not explained, that has to be read as an indictment of any Catholic woman who would dare to limit her fertility. Furthermore, the document prescribes for all women one of either two sexual states: married biological motherhood or virginal spiritual motherhood. Because Jesus' mother, Mary, managed to embody both states, motherhood and virginity, she has been held up by John Paul II elsewhere as a sterling example of womanhood, showing "how two dimensions, these two paths in the vocation of women as persons, explain and complement each other." No such paths, of course, have ever been laid out by the Vatican for men.

In the 2004 document, the Vatican charges that women who dare defy its twin imperatives are causing the downfall of the traditional family. Such women "call into question the family in its natural, two-parent structure of mother and father," said the document. It also accused such women of making "homosexuality and heterosexuality virtually equivalent." But how do women do this? It is never explained, but again, one must assume women do so by defying their fundamental natures, as defined by the Vatican; by joining the multitudes of nonvirginal, nonmothering, sexually active Catholic women; by using birth control and ending pregnancies they know they cannot manage; and by heading up alternative, including lesbian, families—all out of the Vatican's control. If the hierarchy cannot control women, then it cannot control the Catholic family, and if it cannot control the Catholic family, then—particularly in the wake of the priest child sex abuse scandals—it loses any remaining claim to moral authority.

This, of course, isn't anything new. Indeed, that recent Vatican letter is part of a long history of Church attacks on anyone who would step out of its approved sexual and marital paradigm—attacks that Catholic women reformers continue to meet head-on.

REDEEMING THE CHURCH'S "SEXUAL OUTLAWS"

Because of the Church's narrow sexual teaching—that the only acceptable sexual activity is between a married man and woman open to procreation—the vast majority of sexually active Catholics qualify as sexual outlaws. They range from teens to the elderly, from homosexuals to single heterosexuals, from married couples using birth control to remarried Catholics who never got an annulment. Paradoxically, the Church has aided, abetted, protected, and made excuses for its *real* sexual outlaws: priests who sexually abuse and exploit children and adults and the bishops who cover for them. This state of affairs has stirred Catholic women reformers to action; they see the day as long overdue for a new, intelligent, and humane Catholic sexual ethic.

Fighting for the Rights of Catholic Homosexuals

Surely, homosexuality is the only issue in the Roman Catholic Church that is almost as controversial as abortion. In the Church's sex abuse crisis of 2002, demands to address priest pedophilia became demands to oust and exclude from the ministry all gay Catholic priests—as if homosexuality itself, any more than heterosexuality itself, causes the sexual abuse of children. Then, in the years that followed, the Church unleashed its homophobia on lay homosexual Catholics, denouncing the movement to legalize gay marriage in terms better applicable to the most heinous outrages of the modern world.

"Those who would move from tolerance to the legitimization of specific rights for cohabiting homosexual persons need to be reminded that the approval or legalization of evil is something far different from the toleration of evil," wrote Pope John Paul II, in a public statement about gay unions. In a stunning example of moral chutzpah for the leader of a Church whose own ministers have done true violence to children, the pope went on to say that "allowing children to be adopted by persons living in such unions" was "gravely immoral" and "would actually mean doing violence to these children." The document declared homosexual acts to represent "a serious depravity." So slippery was the hierarchy's slope on the subject of homosexuality that Brooklyn Bishop Nicholas DiMarzio declared publicly that if gays were allowed to wed, society soon would be opening the door for people to marry their pets.

In the Catholic Church, one woman above all others has demonstrated a relentless devotion to including gays and lesbians in the Church family. She has also, in one lifetime, illustrated the stubborn commitment of Catholic women to equality, sexual justice, and a welcoming Church, in the face of the hierarchy's repeated attempts to undermine, indeed destroy, that commitment. She is Sister Jeannine Gramick.

An Advocate in Action

A slightly built woman of sixty-two, with reddish-orange hair and eyes that can conjure the color of the Caribbean on a sunny day, Gramick lives in the headquarters of New Ways Ministry, in a three-story, white-sided house outside of Washington, D.C. Awards, citations, and pictures hang on the first-floor wall. There is a full-page *New York Times* ad reading "A Catholic Pledge to End Violence Against Lesbian and Gay People," with rows and rows of signatories. There are laudatory plaques from nationwide chapters of the Catholic homosexual rights group, Dignity, as well as a "Lifetime

Achievement Award" from Dignity USA. Most interesting to me is a beige-matted, sepia-colored photograph in a gold oval frame. Standing, in a long, flowing dress reminiscent of the 1920s, holding a parasol, is Father Robert Nugent. Seated in a black tuxedo and top hat, right hand resting on a shiny black cane, is Sister Jeannine Gramick. They took the picture on a lark, in a playful photo studio on North Carolina's Outer Banks—two earnest, daring, cross-dressing sexual outlaws.

The first-floor dining room serves as New Ways' conference room, while upstairs are the organization's offices as well as Gramick's bedroom (where her eleven-year-old cat, Rosie, is sprawled out on her bed) and a room inhabited by Sister Alice Zachman, Gramick's dear friend and a longtime social activist, who works on behalf of torture survivors. Living at New Ways is one of Gramick's provocations, taken in the interest of her ministry and in defiance of a Vatican that has done everything short of taping her mouth shut and throwing her in jail to curb her power and influence. Actually, Vatican police did once detain her, in 1994, along with several other Catholic sisters. Then cochair of the feisty National Coalition of American Nuns, Gramick had traveled to Rome to protest yet another all-male Vatican Synod on Religious Life, carrying the now infamous banners that read: "No speaking about us without us." The Vatican freed the nuns but kept a banner.

Gramick's advocacy for the rights of gays and lesbians in the Church began in the early 1970s, when she was a young nun with the School Sisters of Notre Dame in Philadelphia. There, she met Dominic Bash, a gay man who reluctantly had given up on the Catholic Church of his childhood. There were others, too, and before she knew it, Gramick was rallying priests to serve home Masses for gay men who had been alienated from the Church, who began, on the strength of her acceptance of them, to return. Her activities attracted press attention, which brought some "fan mail," including a letter from Salvatorian priest Father Robert Nugent. Gramick contacted Nugent, and thus began a thirty-year association that would birth a historically groundbreaking ministry to homosexuals from within the Catholic Church—New Ways Ministry. Founded in 1977, New Ways defines itself as a national effort "to build a bridge of justice and reconciliation for gay and lesbian people and the wider Church."

Gramick's and Nugent's workshops and pastoral welcome brought thousands of homosexuals out of the Catholic closet and into the arms of the Church. The two talked about homosexuality in the context of biology, sociology, history, culture, and the family. They talked about biblical denunciations of homosexuality, noting that they have been questioned on many grounds, from their relative rarity in the Bible to a contention that the primary subjects of those passages were rape, prostitution, or pedophilia, not homosexuality. They covered theology, sharing the alternative perspectives of lesbian/feminist theologians who were "exposing the limitations of a procreative sexual ethic and . . . suggesting instead an ethic based on mutual relation," and psychology, this not long after the American Psychiatric Association ended its classification of homosexuality as a mental illness.

They set out to convey the full range of Church teachings. They talked about the immorality of homogenital behavior, described in Church documents as "intrinsically disordered," but also the newer teachings, such as the "moral neutrality of a homosexual orientation" (that is, while the behavior is immoral, the orientation is not). They talked about the necessity for pastoral care and "the immorality of prejudice and discrimination" against homosexuals. Gramick and Nugent wrote and published, too. After many articles and handouts, their first book, *Building Bridges: Gay and Lesbian Reality and the Catholic Church*, came out in 1992. By then, they had conducted seminars in more than 130 of the country's 169 dioceses and archdioceses.

In the mid-1990s, Gramick and Nugent encouraged the U.S. bishops to create a document on homosexuals. They contacted Bishop Thomas Gumbleton to introduce the project and gathered together what Gramick called a group of "gay-friendly bishops." Nugent provided consultation to the bishops. Published in 1997, *Always Our Children* was a clear validation of the need to minister to homosexuals with sensitivity and love, and to help their parents to understand homosexuality and accept their homosexual children with love. The document in tone and content was deeply pastoral.

By the turn of the twenty-first century, what began as a fledgling outreach to homosexuals in the Catholic Church had grown significantly. The National Association of Catholic and Diocesan Lesbian and Gay Ministries, founded in 1994, reports that some 149 parishes and thirty-one dioceses have had ministries to Catholic homosexuals. In 2003, the National Association, with the Center for Applied Research in the Apostolate, surveyed twenty dioceses and fifty-two parishes with special ministries to lesbian and gay Catholics active in the United States; they found a range of services offered, from referrals to counselors and spiritual directors to outreach, to support groups for homosexuals and their families.

The Battle Begins

It would be nice if that were the end of the story. It isn't. The two paid an enormous price for their ministry. From the late 1970s through the 1990s, they were the targets of a Kafkaesque investigation by local Church authorities and the Vatican that eventually brought the full wrath of the hierarchy down on their heads. The first sign of trouble came over Gramick's ministry not to homosexual lay Catholics but to lesbian nuns. Gramick had been asked by a group of nuns to hold a retreat on homosexuality in the spring of 1979. When Washington, D.C., Cardinal William Baum learned about it, he notified the Vatican.

"They assumed that if lesbianism was going to be discussed, and if there were lesbian nuns, then they were sexually active," says Gramick. No doubt, some were; some were not. But that wasn't the point of the retreat. Gramick says that it was about sexual identity, about lessening feelings of isolation for women who felt "I'm the only lesbian nun there is in the world." It was intended to give the nuns what any gay person needs: support. The Vatican pressured Gramick's superior general to cancel the retreat, and the superior general conveyed that concern to Gramick but took no step to stop it. Gramick held the workshop, as planned.

The sexuality of priests and nuns remained an important aspect of the ministry of Gramick and Nugent, whose willingness to take on that controversial subject proved prescient. "Many in the Church suspect that the great resistance to and suppression of any development regarding sexual issues within some segments of the hierarchy are due to an unwillingness to face homosexuality within its own ranks," she wrote in *Building Bridges*. "It is a time in which the faith community may be ready and willing to listen to and actually hear the voices of lesbian and gay ministers in our midst . . . to acknowledge that [they] are an important and vital part of the Body of Christ."

Their workshops continued, and so did the monitoring. Pressure from the hierarchy built on their religious communities to take action against Nugent and Gramick, which they did, by conducting internal examinations of their work. The communities found nothing untoward and continued to support Gramick's and Nugent's efforts. But the controversy was fast becoming public, with statements of support coming from homosexual advocacy groups as well as hundreds of gay and lesbian Catholics.

Those early years of investigation came to a head in 1984. Upon Vatican order, Gramick and Nugent were compelled to "separate themselves totally and completely from New Ways Ministry." They were also forbidden to participate in any program unless they made clear "that homosexual acts are intrinsically and objectively wrong." Gramick did separate herself from New Ways "legally," she told me, taking her name off the letterhead and ending her official role. But she admits, "I was still involved in the sense, behind the scenes, supportive, and afraid that that would be discovered." That year, Gramick moved from the D.C. area to Brooklyn, New York, the only diocese on the East Coast that was willing to accept her. She continued her ministry to the homosexual community, writing, speaking, and working with various groups, but not as a member of New Ways Ministry.

The investigation escalated. In 1988, the Vatican Congregation for Religious and Secular Institutes established a special U.S. commission to study the writings and ministry of Gramick and Nugent, headed by Bishop Adam Maida. Years later, in October 1994, the Maida Commission released its report, which astonished Gramick and Nugent in its "harshly negative tone." It conceded that the two were "engaged in an important and needed ministry" and were "to be commended" for their "courage," "zeal," and "love and compassion." But it went on to a scalding criticism with ominous overtones. The commission condemned the two for not promoting the position that homosexuals should work at changing their orientation, a profoundly outdated notion that the commission said could be bolstered by the "considerable testimony among evangelical Protestant communities." They chastised them for criticizing Church statements, including the 1986 Vatican *Letter on Pastoral Care of Homosexuals*, which had been publicly decried by many observers when it was issued. In one breath, that 1986 Vatican letter said it was "deplorable" that homosexuals have been "the object of violent malice in speech and in action." But in the next, it rationalized that violence, saying that if people "condone" homosexual activity or promote civil legislation, no one should be "surprised when . . . irrational and violent reactions increase." And the commission made recommendations about the fate of Gramick and Nugent but refused to tell them what those recommendations were.

Gramick and Nugent were deeply disturbed by the secretive nature of the process and said so in their official response to the commission report. They said that they had been forbidden to examine all the letters sent to the commission and passed on as evidence to the Vatican; that they were never allowed to see the full text of the charges against them; that one of their presentations had been secretly taped, and only unfavorable reviews used; and that they offered as evidence some 250 letters—including letters from fifteen bishops—in favor of their ministry but the commission refused to make those letters part of the official record.

Just a few months later, Gramick and Nugent received more questions in writing. Their answers, unbeknownst to them, were reviewed by the echo of the Inquisition, Cardinal Ratzinger's Vatican Congregation for the Doctrine of the Faith, to whom their case had been secretly referred. Based on its examination, the CDF prepared a medieval-sounding document called a *contestatio*. It held that there were "erroneous and dangerous positions" in their books—*Building Bridges* and *Voices of Hope*, their new book—"which have caused grave harm to the faithful." It demanded that the two respond.

On some points, Gramick was conciliatory. She agreed to provide a fuller understanding of the Church's teaching on conscience, ensuring that people carefully consider the Church's teachings so they can avoid forming what is called an "erroneous conscience." But on other points, she held firm, again defending her refusal to focus only on the Church teaching on homosexuality as an "objective moral evil." She said that her intent had always been "to present the full range of the Church's teachings," which included its commitment to pastoral care for homosexuals as well as its condemnation of homosexual acts. Earnest as her response was, it did not do the job. Gramick reported that the CDF was not satisfied "because I never disclosed my personal convictions with regard to the Church's teaching on homosexuality." In other words, the CDF wanted her conscience on a plate.

Gramick and Nugent were ordered by the CDF to sign a "Profession of Faith." Conjuring up images of the demonized Galileo, it demanded "a submission of will and intellect" to the Church's teaching about homosexuality, saying that "homosexual acts are always objectively evil . . . acts of grave depravity . . . intrinsically disordered." Nugent tried to temper the castigation of homosexuality, then signed, but that failed to satisfy Ratzinger. Gramick refused to sign at all.

The response of the CDF was ruthless and unequivocal. In 1999, it condemned both Gramick and Nugent for their ministry and their writings in a published *Notification*. It declared their approach, with its "errors and ambiguities," to be inconsistent "with a Christian attitude of true respect and compassion" and to "have caused confusion among the Catholic people" and "harmed the community of the Church." The punishment: a permanent prohibition from any pastoral work at all with homosexuals.

The *Notification* shocked the gay community, the progressive Church reform community, and some prominent moral theologians, including Boston College's Lisa Sowle Cahill. Writing in *America,* she conceded that Nugent and Gramick might have "downplayed" one aspect of Church teaching about homosexuality—"the intrinsic evil and immorality of homosexual acts." However, she wrote, by focusing on that fact, "the Vatican policy on conformity and dissent has certainly taken a novel turn." Instead of silencing people for contradicting doctrine or "noninfallible teaching," the Vatican was now silencing them for "staying within Church teaching" but not doing it well enough. She characterized the new "bottom line of orthodoxy" as requiring "not only theologians but pastors to present a range of specified points of Catholic teaching around a given issue, and to do so in precisely the language prescribed by Vatican authorities."

Gramick was devastated by the *Notification*. She went on a retreat to decide how she would respond—and came back fighting. In a public statement, she wrote that she would comply with the order to end her ministry to homosexuals, saying she felt it was "more beneficial to minister on their behalf with the blessing of Church leadership than without it" but made clear how painful complying with that prohibition would be, like "living with . . . a heavy weight on my heart and soul."

Then she did everything she could to rally support to have the CDF decision "reconsidered and, hopefully, ultimately reversed." In speaking engagements all over the country, Gramick talked about the issues as she saw them: "the negative effects of the CDF decision on lesbian and gay persons and their families," the problem of the hierarchy's "interference in the internal affairs of a religious congregation," and "the role of a public minister, the right of privacy of conscience . . . the fragility of human rights in the Church, [the need for] fair and just procedures and penalties, nonviolent resistance to unjust Church laws, identifying the central teachings of the Church, and . . . legitimate dissent." She asked people to write in support of her, and they did. "I have a suitcase full of letters," she told me, which she estimates at well over a thousand. And Gramick was told that her superiors in Rome also received thousands of letters, by which they were quite "annoyed."

The Vatican attacked again. They wanted her completely silenced. But unlike past years, when Gramick had the support of her religious community, this time, the superiors of her religious order complied. In May 2000, Gramick was called to Rome to receive a list of "obediences." "They said that I may not speak about homosexuality. I may not speak about the investigation, either privately or publicly. I may not criticize the Magisterium. I may not incite the faithful to protest in my regard." A similarly worded silencing order was given to Nugent. He complied; Gramick refused. She issued a statement that read in part: "I choose not to collaborate in my own oppression."

Gramick went on speaking, despite grave warnings from her order. She knew that soon, if she persisted, she would be expelled from her spiritual community for more than forty years. On the eve of the 2001 Call to Action conference in Philadelphia, where she would receive their annual Leadership Award "for dedicated work with gay and lesbian people and their families, which has enlightened and healed many and opened up an important dialogue in our Church," Gramick was preparing to leave the School Sisters of Notre Dame.

Gramick Today

Today, Gramick is a member of the Sisters of Loretto. To her mind, because the silencing order came from the School Sisters, it no longer applies. "I can speak about the investigation and not have any sanctions," she says with a smile. "I can speak about homosexuality. I can criticize the Magisterium. I can incite the faithful to protest. . . . I'm not breaking any vow of obedience, if you want to get legalistic about it." But the Lorettos, too, are under the Vatican's authority. Is she really any safer? "The Vatican could come after them, as the Vatican came after the School Sisters of Notre Dame," Gramick explains. "But I'm safer because the Sisters of Loretto are willing to take more risks."

Indeed, the Sisters of Loretto have been risk-takers for decades, being vocal advocates for women's rights in the Church, for women's reproductive rights, for the rights of homosexuals. Gramick credits the Lorettos' risk-taking spirit to their frontier roots in Kentucky, from which the sisters moved westward in the early 1800s to build their American religious community. But the fact is, Gramick's status in the Church remains much closer to perilous than safe. At any moment, she knows, the other shoe could drop, pressure could be put on the Sisters of Loretto, and she could ultimately be banished from that community, too, needing to find another way to live out the last years of her life.

But Gramick says she has accepted that. Which helps explain her eagerness to bring her name before her nemesis, Cardinal Ratzinger, once again. In 2004, her book *Building Bridges* was published in Italian, and she went on a press junket to Italy. While there, she decided to leave a copy of the book and a note for Ratzinger. He wasn't a complete stranger to her. Gramick had had a chance meeting with him on a plane in 1999, just before the Vatican issued its *Notification* banning her from any ministry to Catholic homosexuals. She had taken a deep breath, gone over to introduce herself, and sat with him briefly. She took the opportunity to make a plea not for herself but for Catholic homosexuals, telling him how deeply devoted she had found them to be to the Church and how much her work with them had enriched her life—an encounter that obviously failed to stay in Ratzinger's hand. In 2004, Gramick scribbled similar sentiments on the note, which she ended up having to leave with his secretary. She never heard back.

At this point, the public recognition Gramick has received may well inoculate her from further backlash. In 2004, Barbara Rick's inspiring and engaging documentary film, *In Good Conscience: Sister Jeannine Gramick's Journey of Faith*, was released, bringing Gramick's story to the world. It had its sold-out U.S. debut at the Walter Reade Theatre at Lincoln Center in New York City before a packed, standing-room-only crowd who gave the two of them—Gramick and Rick, a Catholic, too—a rousing standing ovation.

Today, Gramick wants to see more Catholic ministries to homosexuals and more freedom for gay men and lesbians to come out. She believes that "coming out is the best thing," though she advises that it be done with care for repercussions. Actually, she sees a universal message in overcoming the fear that prevents people from taking this action, and many others that call for courage. "Every one of us, myself included, we're ruled by fear.... We make decisions on the side of caution ... when we should make decisions on the side of boldness. So often, I have seen [that] when [people] make decisions on the side of boldness, the fears they have usually do not materialize. Now, I'm not going to say always. But usually."

Which brings me to ask her, Is she a lesbian? Gramick demurs. She's never said publicly. She feels no matter which way she goes, she can't win. "If you say you're heterosexual, they'll say you really don't know what you're talking about," she explains. "If you say you're a lesbian, they'll say, 'Well, you're just doing this for . . . self-promotion.'" She admits she has been identified publicly as a lesbian and as a heterosexual. "Maybe I'm bisexual," she says coyly. "Maybe I'm transgendered. Who knows?"

Not surprisingly, she believes in a world where sexual orientation wouldn't really matter. "We shouldn't have to ask people what your sexual orientation is. You shouldn't have to ask people what your race is. . . . Who cares? Gender is a little harder," adds the woman who posed as a dapper man in that Outer Banks photo. "But I would love it if men and women were indistinguishable, if we were all androgynous." She sees all of those characteristics as "accidental." They are not, she says, "the essence of what it means to be a human."

Carrying On the Campaign for Change

True acceptance of homosexual Catholics in the Church demands that the hierarchy stop speaking out of both sides of its mouth. Official policy discourages discrimination based on a "homosexual orientation" but condemns homosexual acts as "intrinsically disordered"—what one reformer described as the "it's okay to be a dog, just don't bark" school of thought. But in 2003, in its statement denouncing gay marriage, the Vatican stepped up its heated rhetoric calling the homosexual "inclination" itself "objectively disordered"—which, when compared with "intrinsically disordered" homosexual acts, sounds like a distinction without a difference. Change also requires that the Church base its sexual teachings on science, which since the 1970s has concluded that homosexuality is not a psychiatric pathology. And change demands what Gramick and Nugent called for decades ago: a Church willing to face the high representation of homosexuals in its clergy and religious life, and to recognize the contributions of homosexual clergy and religious to building the American church.

Today, many progressive reform groups support this struggle. Though barred from most parish venues, Dignity remains active, with 2,300 members and forty-six chapters around the country. While gay men have predominated among its membership, women are active, too; in 2002, national Dignity had both a female board president and a female executive director. Other supporters include the Leadership Conference of Women Religious; the National Coalition of American Nuns; Call to Action; gay and lesbian associations on college campuses; and newer groups like Rainbow Sash in Australia, whose members wear rainbow sashes over their shoulders to Mass and to Communion to signify that they are gay and lesbian Catholics "who embrace and celebrate our sexuality as a sacred gift from God," and Soulforce, which describes itself as an interfaith movement that is committed to ending spiritual violence against homosexuals that is perpetrated by religious policies and teachings.

As for Catholics in general, there is a very long way to go, but there are also some signs of change. According to one poll, 61 percent of Catholics believe that homosexual behavior is wrong, but 83 percent believe that discriminating against homosexuals is wrong, too. Breaking Catholics down by age group, another study reported that among pre-Vatican II Catholics (sixty-three and over), 69 percent thought homosexual acts were always wrong, but among Vatican II and post-Vatican II Catholics, only 39 percent did.

When compared with people belonging to other major Christian religions in the United States, Catholics—even in the heat of the gay marriage controversy in 2003—had a more favorable opinion of gay men and lesbians (46 percent) than any other group. An estimated half of Catholics would allow homosexual men to become priests, as long as they remain celibate. And more than half (54 percent) of white Catholics think that homosexuals cannot change their orientation, a belief strongly associated with positive opinions about homosexuals—which may bode well for a continued change in Catholics' views.

In the meantime, following in Gramick's footsteps, lesbian and gay Catholics continue to minister in the Church and occupy the pews, gradually changing minds and hearts. In May 2003, Kathy Itzin, forty-two, director of elementary faith formation at Saint Joan of Arc Catholic Church in Minneapolis, was among six people nominated to receive an "Excellence in Catechesis" Christian teaching award. An employee of the archdiocese of Minneapolis and Saint Paul for twenty-one years—fifteen at Saint Joan's—Itzin was described in the diocesan article announcing her award as "compassionate," "collaborative," and given to "exemplary pastoral responses."

But the day after word of her award became public, a right-wing Catholic group called Catholic Parents Online reportedly phoned Archbishop Harry Flynn. Their mission: to out Itzin as a lesbian mother raising four children (three adopted) with her same-sex partner and to demand that the award be withdrawn. Flynn caved under the pressure. In a statement, he said that he had rescinded Itzin's award because he could not "condone the homosexual lifestyle" or her "actions," that is, "homosexual activity" and "artificial insemination," which he declared to be "contrary" to the "consistent teaching of the Church."

It was a devastating moment for Itzin. Very soon, however, a wave of support for her and her family rose throughout the parish, the diocese, and other parts of the country. On the night that the awards were to be bestowed, several hundred people, including one of Itzin's sons, picketed the event held at a local church. In the weeks that followed, the family received tremendous support from pastor Reverend George Wertin, who told the *Star Tribune* he "disagreed totally" with Flynn's decision, saying, "Kathy is a wonderful, respected member of our parish community and staff." Even Archbishop Flynn agreed to meet with Itzin, a meeting she described to me as extremely warm and productive, where he was "really open and really kind."

And on Pentecost Sunday, Saint Joan's parishioners gave Itzin their own award, which Itzin says "meant more to me than anything else." She described it as a beautiful framed parchment with a background of pictures of kids from the parish. It read: "We, the community of Saint Joan of Arc, offer this award . . . to our beloved Kathy Itzin, whose teaching and example have shaped and inspired the faith formation of our children and families. Presented with gratitude."

Reverend George Wertin preached on "how the Spirit blows where it will and calls us to change in the Church and the world," reports Itzin, while former pastor Harvey Egan, "the great renegade priest who made Joan of Arc a place known for social justice, peace, and activism in the world, surprised everyone by coming." At eighty-nine, he hobbled up to the altar on a cane, with an escort, to declare his love and respect.

Supporting the family in the weeks that followed were parishioners from ten area parishes (including Saint Joan's) that formed a network of what they called "Inclusive Catholics." The network officially joined local parishes that welcomed lesbians and gay men and committed to working together for justice, understanding, and change. One year later, according to press reports, Flynn concelebrated a Mass with five of the seven U.S. cardinals (one of whom had previously turned homosexual Catholics away at the altar); members of Soulforce, Rainbow Sash, and Dignity, all wearing rainbow-colored sashes, received the Eucharist.

Making Marriage Disappear: The Plight of Divorced Catholics

Catholics who remarry without an annulment of their first marriage are not subject to the storm of public disdain from the Church hierarchy for their behavior as are Catholic homosexuals. But they, too, are sexual outlaws. A Catholic who plans to remarry must have her first marriage annulled—which is the Church's way of making it magically disappear. An annulment is a finding, according to canon law, that the marriage was canonically invalid from its inception, that the sacrament of marriage never took place. A Catholic woman who fails to have her first marriage annulled and remarries—which would have to be done outside of the Church—is forbidden to receive the Eucharist and declared by the Church to be an adulterer. The only way around this is if she and her new husband commit to not having sex. As for the children of an annulled marriage, though there is great concern that they become illegitimate afterwards, in fact, in Church law, they do not.

Pope John Paul II's signals about those Catholics who divorce and remarry outside of the Church and without an annulment have been mixed, calling for compassion one moment and censure the next. He has written that such Catholics remain members of the Church, and that pastors should help them "to experience the charity of Christ," receive them "with love," and invite them to participate in Church life. But he has consistently ordered priests to withhold the Eucharist from those divorced and remarried without an annulment, who he has included among those who "obstinately persist in manifest grave sin." He has urged Catholic lawyers to refuse to handle divorces, arguing that lawyers "must always decline to use their professional skills for ends that are contrary to justice, like divorce." And he has chastised the faithful, including canon lawyers, for failing to take seriously enough the "indissolubility" of marriage.

In addition to banishment from the Eucharist, the Church hierarchy in recent years has used remarriage without annulment as grounds for firing remarried Catholics from their jobs. "All over the country, Catholics who remarried without an annulment and who work in church jobs or teach in church schools or CCD [religious education] programs are being fired because they are not 'following the teachings of the Church," observed Charles N. Davis, a member of the advisory board of Catholics Speak Out and resident divorce expert.

Indeed, according to press reports, Angel Meacham, a fifth-grade teacher for seventeen years at Saint Joseph Elementary School in Crescent Springs, Kentucky, was fired in 2003. That was after diocesan officials received a letter informing them that she had remarried in a Presbyterian Church, apparently after running out of patience with the annulment process. After losing her teaching job for the same reason, Vicki Manno sued Saint Felicitas Elementary School in Euclid, Ohio, its principal, and its parish pastor.

Divorce wasn't always forbidden in the Catholic Church, nor was marriage always a Church affair. Jesus denounced divorce, but feminist historians and advocacy groups like FutureChurch have their own reading of that action. Hebrew women in Jesus' time had no right to inheritance and could be divorced for just about anything, "from burning dinner to adultery," wrote FutureChurch executive director Sister Christine Schenk. By contrast, women were forbidden from divorcing their husbands. For women, divorce was an unrivaled disaster; they could barely survive alone. "Seen in this light, Jesus' proscription of divorce is markedly protective of women," observed Schenk. In the first three centuries, marriages were family matters that had nothing to do with Church officials. For a time, adultery remained grounds for divorce and remarriage—for men, not women. But in the twelfth century, the Church declared that the marriage bond could not be dissolved, and in the thirteenth century, marriage became a sacrament. Church annulments became the trap door out.

Before the Second Vatican Council, the grounds for annulments were narrow. A Catholic had to prove a lack of consent, essentially due to insufficient use of reason—e.g., by virtue of insanity. After the Council, however, the Church greatly expanded the psychological grounds for annulments, under the rubric of "lack of discretionary judgment about the rights and obligations of marriage," to include many new grounds, from alcoholism to emotional immaturity. With that change, the number of annulments in the

United States skyrocketed, from 338 in 1968 to nearly 47,000 in 2002. Almost 70 percent of the nearly 70,000 annulments given worldwide were given in America, to Pope John Paul II's continuing dismay. Still, Catholics are thought to divorce at rates similar to other Americans—some 40 percent of all marriages, while an estimated 80 percent of them reportedly never apply for an annulment.

The Annulment Experience

Some say that an annulment can be healing, "the final letting go" of a failed marriage, reports Irene Varley, executive director of the North American Conference of Separated and Divorced Catholics. And some dioceses have developed programs that she thinks insert a much-needed pastoral hand into a decidedly cold, legalistic, and invasive process. Varley herself worked for seven years in divorce ministry and bereavement as an employee of the diocese of Columbus, Ohio, running retreats and support groups, including one for remarrying Catholics—whether they planned to seek an annulment or not.

But for many more Catholics, the annulment process is gut-wrenching. It requires a public airing to a local tribunal of the gory details of one's life—from the pitfalls in your parents' relationship to the conflicts you had while dating your husband, to any "unusual incidents" that happened on your honeymoon. It is an even harder process for some now, since the same clerics who have been at the center of the clergy child sex abuse crisis serve on those tribunals, the very people, says Varley, who are "telling us we should stay in our marriages." And the process has been highly politicized in years past with swift and easy access to annulments for the monied and powerful.

When the late Robert Kennedy's son, former congressman Joseph Kennedy, decided he wanted to remarry in the Church, he first had to obtain an annulment of his marriage to Sheila Rauch. For Rauch, an Episcopalian, that meant that one day, out of the blue, she received a formal notice from the Boston archdiocese. To her shock and dismay, she saw that her husband of nearly thirteen years had, unbeknownst to her, initiated annulment proceedings. While she had accepted that her marriage was over and in fact was the one who filed for divorce, she was not ready to declare that the marriage had never *existed* in the eyes of God or the Catholic Church. Rather than simply signing on the dotted line, Rauch fought back. She wrote a book, *Shattered Faith*, in which she took not only her former husband but the Catholic Church to task over the annulment process.

Rauch went beyond publishing her book to join forces with Catholic activist Janice Leary to help other women who are against annulment to fight the process. Together, they founded an organization called Save Our Sacrament, or SOS. "We're doing the reform one woman at a time," says Leary, a longtime progressive Church reformer with Call to Action, Massachusetts Women-Church, and Voice of the Faithful. SOS provides women who are "respondents" in annulment cases—that is, their husbands have petitioned for annulments—with support, education about the annulment process, and guidance about the respondens' rights, including the right to advocates, to read evidence, to have tribunals hear their cases, and to appeal the local decisions to Rome.

Leary says she has seen enormous damage come from the annulment process. She tells of a man who used postpartum depression to suggest a wife's insanity, which the tribunal then passed by a psychologist, who then declared the woman insane, never having met her. Leary has seen "respondents who have died, heartbroken . . . because of what they had to go through, what their husbands wrote, and then what the Church did." And she reports being "truly scandalized" by the fact that if a Catholic wants to wed a formerly married non-Catholic—say, a Protestant who married another Protestant in a valid ceremony—the Church must declare that original non-Catholic marriage annulled before a new marriage can be celebrated. Leary describes a case where "the Catholic had had an obvious public affair with a Protestant, and the Catholic tribunal annulled the Protestant marriage so that the man could marry the woman he had an affair with. It's so ludicrous that, if it wasn't so serious, one could only laugh."

Leary fought an effort to have her own marriage of several decades annulled. She thinks the process was hardest on her three daughters, which is often the case. "My tough lawyer daughter broke down and sobbed when she heard about the annulment," she says. "Divorce was something they could deal with. But annulment is devastating for children in a way that I hadn't even expected." Because of Leary's appeal, her ex-husband's petition for an annulment has been on hold for years.

The Catholic reform organizations working on divorce—from Save Our Sacrament to Catholics Speak Out, from the North American Conference of Separated and Divorced Catholics to the Association for the Rights of Catholics in the Church—publicize a little-known and controversial route to restoring to Catholics the most cherished right they lose if they remarry without an annulment: the right to receive Communion. They propose the "internal forum solution." The Church's annulment process depends on a governing body to reconcile the remarried Catholic to the Church and is therefore an external forum process. By contrast, the internal forum solution is an interior process that relies on the Catholic's own conscience. A divorced and remarried Catholic essentially searches her conscience to attest to the morality of ending her first marriage and entering a second marriage, and usually she does this with the support of a pastoral advisor or a priest in confession. Finding a Catholic priest to act as advisor today can be a challenge. For years, internal forum as applied to divorce and remarriage had vocal clerical supporters, but today that is much less so. At the 1980 Synod of Bishops, Pope John Paul II "effectively ruled out" the use of internal forum as an alternative to annulment.

As a divorced and remarried Catholic myself, I was disappointed to see how little of the reformers' energies are going into this area of Catholic Church reform. While there are, as noted, reform organizations working on the plight of the remarried, and women figure prominently in most of them, this is definitely not an active area of change-making. Websites are stale. Money is scarce. "As far as I can see there is not much of an audience for reform," concludes Catholics Speak Out's Charles Davis. "The bishops think they have a marvelous thing going in the annulment process and in its 'great compassion,'" he says. Indeed, on the one hand, the cost is relatively low, generally from $500 to $1,000, with allowances made for those who cannot pay, though there is an additional similar charge for appealing a local tribunal's decision to the Vatican. On the other hand, observed Davis, "The laity don't want to put up with the hypocrisy of it all, much less have to relive the bad memories of their former marriage that the annulment process demands."

After remarriage, some Catholics do walk away. But many others stay after taking vows in a non-Catholic service. They have found their own way to be Catholic by listening to their own consciences, even if they've never heard of internal forum. Indeed, nearly three-quarters of Catholics disagree with the Vatican that Catholics who remarry without an annulment should not receive Communion. Under 21 percent of Catholics—even fewer women than men—look to the Church hierarchy for moral guidance about whether they should remarry after divorce without an annulment.

Today, many remarried Catholics have blended into parish life. They may find a parish where they can receive Communion without telling the priest, which Charles Davis characterizes as the "don't ask, don't tell" approach. Some priests know about the remarriage but give the couples in their congregations Communion anyway. One particularly feisty Church activist read over the annulment papers provided by her diocese and concluded that she could never go through such an ordeal. She had her second marriage ceremony in an Episcopal Church, then returned to her Catholic parish. When she asked her priest if he would still give her Communion, he was hurt that she would even ask but also declared that it was a moot point. "In your case," he told her, "I wouldn't deny you Eucharist because you'd just reach in and take it." Other Catholics have been welcomed into the ranks of small faith communities and intentional Eucharistic communities within the hierarchical structure, while still others belong to small independent Catholic communities headed by married Catholic priests or Catholic women.

Dangerous Marriages

While many U.S. Catholics get annulments (an estimated 80 percent to 90 percent of all accepted cases are approved) or simply remarry anyway, remaining quietly in the Church, in very Catholic countries, conservative attitudes toward divorce can force women to remain in marriages that are dangerous to their lives. In fact, in some poor countries in Latin America,

Africa, and Asia, annulments are completely inaccessible to people because there are no diocesan tribunals at all.

When Catholic high school chaplain Katie Bergin was in Chile working as a Catholic missionary, one of her friends went to confession and talked about her abusive marriage. Instead of sympathizing, reports Bergin, the priest told her she couldn't "get divorced because divorce is a sin." In 2000, when Bergin had this experience, divorce was illegal in Chile, which by no coincidence, says Bergin, "is a very Catholic country." In fact, not until 2004 did Chile finally legalize divorce. "I don't know how to make sense of a Church that is supposed to be about 'life' telling women to stay in abusive marriages," she says.

The roots to that thinking run deep. Augustine, bishop in the early Church and esteemed Church father, offered his mother as a model of holiness and Christian fidelity for other women to follow. He wrote: "When she had arrived at a marriageable age, she was given to a husband whom she served as her lord . . . as he was earnest in friendship, so was he violent in anger; but she had learned that an angry husband should not be resisted, neither in deed, nor even in word. . . . [W]hile many matrons, whose husbands were more gentle, carried the marks of blows on their dishonored faces, and would in private conversation blame the lives of their husbands, she would blame their tongues."

Unfortunately, such notions die hard. To their credit, the U.S. bishops in 2002 released a strongly worded pastoral response to domestic violence against women. A woman put that pastoral in motion; it was written at the urging of Dolores Leckey, then executive director of the Secretariat for Family, Laity, Women and Youth, who took the concerns of women around the country and saw the need for such a pastoral to the bishops. The sympathetic document clearly stated that no abused woman is expected to stay in such a marriage, that she should not hesitate to separate or divorce, and that, once divorced, she should seek an annulment.

But negative attitudes persist toward divorce for Catholic women in those and other circumstances and are too seldom actively challenged by Church leaders. "Divorce is a grave offense against the natural law[,] . . . does injury to the covenant of salvation[, and] . . . is immoral," reads the Catholic catechism. Today, thirty years since the founding of the North American Conference of Separated and Divorced Catholics, Irene Varley still gets calls from divorced Catholics who believe they are not welcome in the Church and have been excommunicated. Actually, their status as Catholics changes not at all after divorce, but they don't know that. "They no longer see themselves as holy or worthy," she says, "especially if it's not made clear from the pulpit," which, she notes, it seldom is.

In 1994, Pope John Paul II added to the confusion. He beatified (the step before sainthood) a woman named Elizabeth Canori Mora, born in Rome in 1774, declaring her to be a woman "of heroic love," "exemplary" as a wife and mother. He said that "amidst a great many marital difficulties," she "showed her fidelity to the commitment she had made in the sacrament of marriage and to the responsibility stemming from it."

Upon closer scrutiny, however, this woman had more in common with Augustine's mother and her friends than with the women to whom the U.S. bishops directed their pastoral. Turns out, Mora's husband "compromised" the serenity of the family, "deceived" the family, "estranged himself" from the family, and reduced it "to destitution," reported the Holy See in its newspaper L'Osservatore Romano. She was also clearly a victim of domestic abuse. "To the physical and psychological violence of her husband, Elizabeth responded with absolute fidelity," reported the paper approvingly. "There are no excuses, conveniences, or interests that can justify any distraction whatsoever to the code of fidelity, which is of love and of total surrender." We are apparently a very long way from elevating a divorced woman, even a physically and psychologically abused woman, to sainthood for her courage in leaving her marriage to save herself.

The Church needs to support couples as they navigate the devastating terrain of ending a marriage, but the annulment process comes nowhere near filling that need. This system turns Catholics away from the Church; bars them from Communion; can throw children from a former marriage into a spiritual limbo—even though they are not technically deemed illegitimate; and demands remarried Catholics whose marriages were never annulled to keep their status secret. It is just one more instance of the Church's denial of sexual reality, just one more compromise to the Church's moral integrity.

Sex and the Unmarried

Single, straight, sexually active Catholics—from teenagers through adults—may be the most invisible of the Church's sexual outlaws. They are supposed to be celibate, but many are not. They are on their own in determining what will be their sexual ethic and in maintaining self-love, self-respect, and a claim to holiness anyway.

In addition to the conscience issues, the sexual struggles of Catholic heterosexual teenage girls today are life-and-death struggles as they face the threat not only of pregnancy but of sexually transmitted diseases that threaten their fertility and of HIV/AIDS. Conservative Christians, including the Catholic hierarchy, have worked tirelessly to withhold both information about and access to contraception from adolescents at the risk of unwanted pregnancy, STDs, and AIDS by promoting abstinence-until-marriage sex education. The federal government supports abstinence programs, funds for which religiously sponsored community-based groups as well as public health departments and public school districts may apply.

Despite the fact that the World Health Organization considers programs that address both abstinence and contraception to be the most effective at influencing teen sexual behavior, these "abstinence only" programs are mandated to teach that sexual abstinence is "the only certain way to avoid pregnancy and sexually transmitted diseases," while excluding entirely any reference to contraception, its efficacy, and its importance, teaching only failure rates. And despite the facts that in the United States, more than half of all high-school students have had intercourse and half of all new HIV infections strike young people under the age of twenty-five, federal funding for these programs continues to increase, to $170 million for 2005. While abstinence is a fine alternative for young women who choose that, their health and lives are in danger when it is treated as their only option. It means that if they choose to have sex before marriage at some point, they are completely ill prepared—emotionally, physically, and spiritually—to protect themselves. There is also tremendous religious, including Catholic, opposition to school-based condom distribution programs even though research shows that young people who complete such programs are no more likely to increase their sexual activity and more likely to use condoms when they do have intercourse than their peers without access to such programs.

Sexually active single women and married women who choose not to be mothers are also sexual outlaws. They are unable, within the context of Catholicism, to make an ethical claim for the way they handle their sexuality. I have lived all my life in these shadows—as a young woman, an unmarried adult, and a married woman who chose not to have children. That last one has been a special challenge. In the Catholic Church, where maternal self-sacrifice is perhaps the ultimate feminine virtue, I have felt at times that I am a walking sacrilege. I have dared to value my life for what it is. I never thought God expected me to create another life in order to deserve this one. And I never thought that my decision about childbearing meant that I could not have physical intimacy in my life.

For many of us, the Church's demand for a life of sexual abstinence is arcane and ignored. But the Church has refused to provide a new sexual ethic to take its place. Catholic women have proceeded without them. Into the void have stepped provocative moral theologians, pointing the way to the light.

9

CATHOLIC WOMEN FOR A NEW SEXUAL ETHIC

Many Catholics around the world long ago rejected the Church's draconian sexual mandates. What is needed today is a new sexual ethic that all Catholics—lay and clerics alike—can abide, that will embrace and not exclude, respect and not condemn, that most importantly, will show faith in the faithful. The question is: What must be done to make the Catholic Church a credible voice on matters of human sexuality?

"We need to unlock the doors of secrecy," says Sister Jeannine Gramick. "[That] means, in the area of sexuality, that Church leaders—bishops—should make no pronouncements on sexuality, probably for decades. And they should engage in listening sessions . . . Listen to the lives of married people. Of single Catholics. Of lesbian, gay Catholics. Of transgendered people. Of priests and nuns. That's right. People who've had abortions. People who have had tubal ligations. They should listen to all of these sexual issues that they're obsessed about.

"But listen to what Catholics really, really think. What is the belief of the Church? To me, the leaders should be articulating the faith of the people. Well, they don't even know what the faith of the people is."

She is not averse to ethical guidelines. "I would also want to be on record as saying, in terms of sexuality, we need to use sexuality in a responsible way. People would be able to do that better if we had more realistic sexual guidelines." The route to those guidelines, in her mind, is these "worldwide listening sessions."

To change teachings, of course, requires revisiting the foundations for those teachings. It is a challenge being undertaken today by many Catholic theologians. One of them is Christine E. Gudorf, associate professor in the department of philosophy and religion at Florida International University, a pioneering and thought-provoking voice for a new Catholic sexual ethic.

Gudorf challenges the Church's grounds for its existing sexual ethic, arguing for a more critical reading of both the Bible and the Church's theological tradition. "We are still teaching a sexual code based in fear of the body and of sexuality," she wrote in her seminal book, *Body, Sex, and Pleasure: Reconstructing Christian Sexual Ethics.*" It is based as well in a belief that women lack reason and "only [possess] the image of God through connection to men." Those beliefs, in turn, contribute to the Church's refusal to grant women full moral authority, to respect what Gudorf calls every human being's "bodyright," which she describes as "a prerequisite for full personhood and moral agency in humans."

Regarding the Bible, Gudorf calls for a willingness to dismiss that which is clearly biased by time and place and not revelatory. Dismissing texts that are deemed not to be revelatory "must be done on a text-by-text basis," she explains. "But it is apparent to all persons not blinded by idolatrous uncritical worship of scripture or theological tradition that both are permeated not only with patriarchy and misogyny but also with antisexual attitudes which are in conflict with the central messages of the gospel."

Gudorf calls for an end to "procreationism"—that is, the belief that sex must be first, foremost, and always open to procreation. For one thing, she says, it "focuses moral attention on the ability of a sexual act to procreate rather than on the dignity and welfare of the actor(s) or the relational context in which the sexual act occurs." She chastises the Church for its rigid emphasis on sexual acts—from homogenital sex to masturbation, premarital sex to contraception. This, she says, has led to "moral minimalism," wherein "virtue in sexuality consists of avoiding specific sexual acts," instead of a focus on "constructing sexual relations, genital and nongenital, which are just, loving" and promote "individual and social growth."

Gudorf argues that procreationism implies that "if a licit sexual act (male-female vaginal intercourse) occurs within the properly sanctioned contractional relationship (marriage), there can be no sin." She sees that approach as "responsible in part for the moral blindness of Christian societies regarding such practices as marital rape," as well as sexual coercion, coital pain, domination, and violence. She calls it "unconscionable that many forms of sexual violence are omitted from most textual and sermonic treatments of sexual sin."

Perhaps most importantly, Gudorf contends that by failing to take into account female biology, the Church got natural law wrong, then used its flawed reading to crown procreation as sex's highest good. She defines natural law as "the unwritten law embedded in all of creation, a law that reasoning humans can discern from observation of creation." Gudorf points out that during a woman's lifetime, very little of the sex she has is likely to be for procreation. Recognizing that, and out of a deep concern for the consequences of runaway population growth, Gudorf argues provocatively for a complete reversal of the Church's current sexual ethic. She calls for "reversing the prevailing understanding that sex is normally procreatively open unless special circumstances require contraception," to one in which "sex is seen as *normally contraceptive* [my italics] so that only very special and consciously selected circumstances justify procreative openness."

She also wants to elevate the value of sexual pleasure in the Catholic ethical system. This won't require much effort, she contends, because God has already done so in the design of women's bodies. What the Church missed in crafting its ethical framework, maintains Gudorf, was the clitoris. "The failure to examine embodied female sexuality combined with the tradition's fear and suspicion of sexual pleasure has led to the tradition's ignoring the existence and significance of the female clitoris," writes Gudorf. The clitoris, she notes, "has no function save sexual pleasure." Furthermore, the pleasure the clitoris brings is not synonymous with sexual intercourse—the only procreative act. Many women cannot achieve orgasm with intercourse. To

Gudorf's mind, that has profound implications for moral theology. "If the placement of the clitoris in the female body reflects the divine will," she writes, "then God wills that sex is not just oriented for procreation, but is at least as, if not more, oriented to pleasure as to procreation." Gudorf proposes "accepting mutual sexual pleasure as the primary purpose of sexual activity," which in turn "requires respect and care for the partner and responsibility for avoiding pain and maximizing pleasure for all affected by that activity."

Moral theologian Sydney Callahan has linked the Church's emphasis on both procreation and "the dangers of lust" to its failure to teach its priests to recognize the "interpersonal dimension of sexuality." She has written that this failure, in turn, has led to the Church's minimization of the sexual abuse of children. "When the psychosexual value of sexuality is not recognized, it is easy to deny the enormity of the damage that sexual abuse can do to a young person's development," she writes. "If abusing priests had been dosing young persons with growth-inhibiting hormones, would the priests have been so easily forgiven and secretly reassigned?"

An Early Framework for New Sexual Ethic

Though more than twenty-five years have passed since it was published, a study commissioned by the Catholic Theological Society of America still has relevance today in terms of developing a new Catholic sexual ethic. Among the five members of the committee who prepared the study, entitled *Human Sexuality: New Directions in American Catholic Thought*, was one woman, Agnes Cunningham. One of only a handful of women ever to receive the society's highest honor, its annual John Courtney Murray Award, Cunningham was also the first woman president of the society (1977–78), a post she held when the book was published.

Building on the thinking of the day's pioneering moral theologians, the study—which was far ahead of its time and engendered great controversy within the society—put forth new standards by which to evaluate moral and immoral sexual behavior. Those standards moved away from a focus on sex acts (e.g., masturbation, premarital sex, adultery, fornication, homosexuality, and sodomy) and on the status of the actors (gay, straight, married, and single). Instead, it attempted to define the "values of a wholesome sexuality."

The study characterized as moral sexual behavior which is "self-liberating," "other-enriching," "honest," "faithful," "socially responsible," "life-serving (not necessary in terms of bearing children)," and "joyous." By contrast, sexual conduct would be immoral if it is "personally frustrating and self-destructive, manipulative and enslaving of others, deceitful and dishonest, inconsistent and unstable, indiscriminate and promiscuous, irresponsible and nonlife-serving, burdensome and repugnant, ungenerous and un-Christlike."

Preeminent theologian Rosemary Radford Ruether applauded this framework in an article on "Sex and the Body in Catholic Tradition" that appeared in *Conscience* magazine in the year 2000. "Such a revisionist view of what makes sex moral or immoral would revolutionize traditional teachings," she wrote. "It would mean that a married heterosexual relationship in which sex is violent, abusive, and exploitative, where the man dominates the woman and exerts his demands selfishly without regard to her feelings and needs, is deeply immoral. By contrast, an unmarried relationship which has qualities of love and mutuality, although somewhat defective in commitment, is nevertheless more moral than an abusive relationship in marriage." Neither the Theological Society nor the American bishops accepted the study's framework as official teaching. But for Ruether "it stands as an important expression of . . . the direction of a revised sexual ethic for Christians (and all people) today."

And Blessings for All

Ruether and many other Catholic theologians and laity also recognize the need for the Church to validate and bless the varied forms that intimate human relationships take in the twenty-first century. In her book *Christianity and the Making of the Modern Family*, Ruether maintains that the "life cycle ceremonies" available to Christians—for example, "monogamous heterosexual marriage for life"—are tremendously inadequate. They either ignore or reject "the more diverse forms of family, such as homosexual couples; the ten to fifteen-year gap between puberty and marriage for many people, occupied by experimental sexual relations; and the breakdown of marriages in divorce."

Ruether argues for new pastoral and liturgical forms, that is, new types of covenants that will join sexual pleasure (*eros*), friendship (*philia*), and loving care of others (*agape*).

She suggests "sexual friendship covenants" for heterosexual or homosexual couples entering into a sexual relationship. These couples, not ready to have children or make a permanent commitment, would take temporary vows that can be evaluated and periodically renewed. These vows would be simple, not involving a public ceremony but bringing "a few friends and mentors together" to confirm and validate the new relationships and be available to help them along.

Young people between their teens and mid-twenties, whose sexual lives Ruether describes as "veiled in lies," would be included among these couples. In Ruether's vision, the ability of young people to enter into these unions would require a radically different approach to teenage sexuality, one that would feature adults helping young people "to learn about their own sexuality in a way that would endorse both sexual pleasure and contraception." It would not be an endorsement of promiscuous sex by any means, but a way to link sex and contraception with friendship and the principles of a responsible relationship.

A second type of covenant—celebrated with a public ceremony—would allow couples to enter into a "permanently committed relationship," after they've had help preparing for their legal, economic, social, and sexual responsibilities. Their commitment would be to "a lifelong effort" toward permanency, in a relationship that might or might not involve children.

The third covenant, at the birth of a child, would be a naming ceremony or baptism. The couple would explicitly vow to "remain faithful to the parenting of the child," to being in the child's life until "the end of their lives," and to do so "regardless of whether they remain in relation as a couple."

Ruether notes that "every covenanting relationship, particularly those envisioned as long-term, must deal with failure." Despite the desperate need for support, she says, this is the moment when many people, particularly parents, "often find themselves most abandoned by community, most isolated and silenced." She argues for healing ceremonies where "partners who are dissolving a relationship work through the transition to new lives, new ways to be friends, ways to continue to parent their children together."

It's Ruether's belief that "once the church is out of the business of being a surrogate for the state in making legal contracts, it can be freed to focus on its more important roles as preparer and blesser of covenants and healer of those who need to move away from covenants that have broken down. . . . It is the church's job to guide the spirituality and ethics of deepening relationship into sacramental bonding and redemptive promise."

Finally, a belief in the duality of body and soul has long pervaded the Church's sexual teachings. That belief made the soiled body the enemy of the exalted soul and women the representation of the soiled body. Despite that history, as moral theologian Sydney Callahan has written, the Catholic Church fundamentally believes "in the goodness of sexual embodiment, the goodness of committed love." After all, at the heart of the Catholic faith is the Incarnation, the willingness of God in the person of Jesus Christ to humbly assume human form, ending the need for any of us to feel alien from the divine. As clinical psychologist and minister Sister Fran Feder has said, "When the Catholic tradition speaks from its best side, it's incarnational, embodied, fleshed. It's sacramental. It's salt and water and ritual and all of those earthly things. . . . When we speak from our best side, there isn't a tradition that could speak more positively about the broad meaning of sexuality."

Many Catholics would love to see the day when the Church is humble enough to approve a truly moral sexual ethic, regardless of what teachings must be undone; when mighty cathedrals all over the world throw open their doors and welcome the outlaws back into the fold; and when the Church trusts all of us enough to make moral choices and walks with us in the real world.

In the interim, I cling as I always have to the law of conscience. I find what I believe is right in the Catholic teachings and principles I was raised with, in the light of my faith in a loving God. To my mind, that law is best described in the *Constitution on the Church in the Modern World*. Here is the way I read it:

"For woman has in her heart a law written by God; to obey it is the very dignity of woman; according to it she will be judged."

10

NEW FOUNDATIONS: CATHOLIC WOMEN RESHAPE THEOLOGY

Moral theology and sexual ethics represent just one branch of theology where Catholic women are blazing trails. In fact, for the first time in Catholic Church history, women represent a substantial number of its most distinguished theologians, and there isn't an area of theology that pioneering Catholic feminist theologians have not touched. "The Catholic community in the United States and worldwide has produced a generation of feminist theologians who have remade theological teaching in every field, from biblical study to historical and systematic theology to ethics, pastoral psychology, and liturgy," writes Rosemary Radford Ruether. Esteemed theologian David Tracy believes that feminism's encounter with religion has produced an "intellectual revolution.... It is in evidence in all the scholarship on religion in every discipline," from doctrine to liturgy to ministry to leadership.

Many women have theology degrees at the master's and doctoral level, and the trend shows no signs of abating. Women represent more than one-third of the students enrolled in the nation's theological schools (Catholic and non-Catholic) and nearly one-third of those studying for a master's in divinity (Mdiv)—the track to ordination. In the last twenty-five years, the number of women in Mdiv programs has increased by more than 225 percent, while the number of men enrolled has fallen by 6 percent. "All of the numeric gain in enrollment in the Mdiv program across the past twenty-five years has been due to the increasing enrollment of women," reports the Association of Theological Schools. Women also make up the majority of students—55 percent—in master's programs in nonordained ministerial leadership.

The number of women studying for doctoral degrees—those officially designated as theologians—has risen, too. Today, 26 percent of the nearly 5,500 students studying advanced theological research are women. At the nineteen theological schools that supply the majority of religion and theology faculty for institutions accredited by the Association of Theological Schools, more than one-third of the students are women. And unlike years past, fully 78 percent of today's women theologians are single or married laywomen, not nuns. There has also been enormous growth in the number of women teaching theology, though there is a long way to go to reach parity: In 1970, only 2 percent of full-time faculty members in Roman Catholic institutions were women; by 2001, 21 percent were—almost exactly the same percentage (22 percent) as at theology schools of all denominations.

Many of these women define themselves as feminist theologians. The classical definition of theology is "faith seeking understanding." To understand feminist theology, advises preeminent feminist theologian Sister Elizabeth A. Johnson, you have to ask: "Whose faith? Whose understanding?" She defines feminist theology as theology "from the perspective of women and women's flourishing." To Ruether, "the critical principle of feminist theology is the promotion of the full humanity of women. Whatever denies, diminishes, or distorts the full humanity of women is . . . appraised as not redemptive," and therefore does "not reflect the divine." Conversely, Ruether has written, whatever does promote women's full humanity is "of the Holy," reflects "true relation to the divine," and is redemptive.

It is a great irony that the Catholic Church, by failing to ordain women, "has created the very conditions under which religious feminism has prospered," noted the late Notre Dame University theologian Catherine Mowry

LaCugna. Indeed, as she observed, "most feminist theologians are Roman Catholic." Among them are the very founders of feminist theology: Ruether, Elisabeth Schüssler Fiorenza, and Mary Daly.

The marks of women's theological scholarship are everywhere. Feminist theologians have reenvisioned God as not necessarily male or female, but male *and* female or neither. They see the Trinity, in the words of theologian Sandra Schneiders, as "*not* two men and a bird." They have reclaimed Eve as human, not evil, and hold Adam responsible for his own fall. They see Mary, Jesus' mother, as assertive, autonomous, and strong, her decision to bear the Messiah between her and God. They found "Wisdom" in the Bible—"the fashioner of all things"—and she is a woman.

They demand an inclusive Church, where everyone is welcome, and inclusive liturgies, where "mankind" gives way to "humankind." They celebrate the female body and claim women's moral authority. They address the pain of black women and Hispanic women at the hands of the Church and society, inventing womanist theology and *mujerista* theology to reflect their struggles and their hope. They reject the Catholic doctrine of the "complementarity" of the sexes, seeing women as fully human and equal to men, active and initiating, and not, as Pope John Paul II sees them, primarily as helpers to men, models of service and self-sacrifice, exemplars of the most passive qualities of "listening, welcoming, humility, praise, and waiting."

Intrepid historians, they look back to the days of early Christianity for what was there and what should have been there, reinterpreting and reconstructing biblical history and women's place in it. Because biblical texts can be hopelessly sexist and mired in the cultural time—used to uphold abominations like slavery—they reject the static notion of "revelation." "God did not just speak once upon a time to a privileged group of males in one part of the world, making us ever after dependent on the codification of their experience," wrote Ruether. "On the contrary, God is alive and with us." They are avowed ecumenists, working exuberantly across faith traditions, foreseeing a reunification of Christian churches, and working toward the day when all the world's religions will coexist equally and peacefully in a truly ecumenical world. And Catholic feminist theologians have come to represent an articulate, alternative voice in the Church, willing to challenge dubious, even dangerous, Church teachings and Church authority. As heads of major Catholic theological societies, they have had the opportunity to multiply their voices a thousand-fold.

That is what happened with Sister Elizabeth Johnson. Author, lecturer, and distinguished professor of theology at Fordham University, Johnson had risen to the presidency of the Catholic Theological Society of America in 1995, when Cardinal Ratzinger, prefect of the Vatican Congregation for the Doctrine of the Faith, declared the ban on women's ordination to be an infallible teaching. Johnson was not happy. Such was the extent of her unhappiness that she would lead the nation's largest and most prestigious association of Catholic theologians in a public and unabashed challenge to Ratzinger and his Congregation for the Doctrine of the Faith. The Church fathers might have foreseen that action had they known what Johnson was *really* thinking when they dragged her, a decade earlier, before a full court of American cardinals to interrogate her about her theology.

Elizabeth Johnson: In the Beginning

Beth Johnson has an impish quality to her. She's short and spry. Her spoken language is a finely honed Brooklynese. Bright-eyed, behind giant old-fashioned glasses, she's ultra-alert. She leans in when she talks to you. She laughs with great fervor. The day I meet her she wears a peach-colored tunic top, brown beads, amber earrings (not dangly), a bangle bracelet, and beige pants. She has brown hair flecked with white, red cheeks, and a creamy complexion, unusual for a woman of sixty-two, which I mean to ask her about.

I find Johnson in 2003 in her windowless basement office on the Bronx campus of Fordham University, getting ready for a move later in the summer. On Johnson's walls are signposts of her life. A framed front page of a New York tabloid blares: "Beth Johnson Gets Tenure: Is the World Ready?" Actually, it's a fake news headline, given to her by her sister—a biological sister, not a member of her order, Saint Joseph of Brentwood. There is a poster of a painting called "The Women Stayed." A gift from her students, it depicts Jesus coming off the cross, surrounded by the women who were there—Mary, Jesus' mother; Mary, wife of Clopas; and Mary Magdalene.

nd the edges of the picture are the images of still other women—Ruth Naomi, Julian of Norwich, Saint Teresa of Avila, Edith Stein, Dorothy Day, and Sojourner Truth.

There is a cover of *U.S. Catholic*, featuring Johnson's book *Friends of God and Prophets*, rimmed with people's faces in individual circles. Based on the Christian concept of the "communion of saints," the book is Johnson's plea for the holiness of ordinary mortals, especially women, instead of "a small group of elite officeholders or canonized saints." Those saints, Johnson has noted, are 75 percent male, with married women the least represented. Johnson's students took the *U.S. Catholic* cover, cut out the faces the magazine published, substituted their own, framed it, and gave it to Johnson—another gift.

And there is her certificate saying that she signed the Madeleva Manifesto, named for Holy Cross Sister Mary Madeleva Wolff. Wolff founded the first graduate theology program for women in the United States in 1944, at Saint Mary's College in Notre Dame, Indiana, at a time when women and laymen were forbidden to study graduate theology at all Catholic universities in the United States. She presided there from 1936 to 1961; in the 1960s, U.S. and European universities finally began to admit women as graduate theology students.

Billed as "A Message of Hope and Courage," the 2000 manifesto declared: "To women in ministry and theological studies, we say: reimagine what it means to be the whole body of Christ. The way things are now is not the design of God." To young women, it said: "Carry forward the cause of gospel feminism. . . . [J]oin us in a commitment to far-reaching transformation in church and society." And to everyone, it said: "We deplore, and hold ourselves morally bound to protest and resist, in church and society all actions, customs, laws, and structures that treat women and men as less than fully human."

On signing the 2000 manifesto, Johnson joined other high-powered signatories: womanist theologian Diana Hayes; moral theologian Lisa Sowle Cahill; Dolores R. Leckey, former executive director of the Secretariat for Family, Laity, Women, and Youth at the U.S. Conference of Catholic Bishops; Monika K. Hellwig, executive director of the Association of Catholic

Colleges and Universities; New Testament scholar Sister Sandra M. Schneiders; and Sister Joan Chittister.

Johnson grew up in Brooklyn, entered the convent as a teenager, and taught grade school for a while, but when renewal of religious life came to her community, she headed for theology school. It was no ordinary school, however. In 1977, she entered Catholic University of America, the only pontifical university in the United States, chartered by the Vatican and under the pope's jurisdiction.

During her years of graduate study, Johnson never had a woman professor, never read a book by a woman, and was often the only woman in a class. Very few female theology students were there at the time, but a handful of them, about ten, formed a study group. "We were the WITS," says Johnson with a chuckle, "for 'Women in Theology.' We even had T-shirts made." Together they read the early feminists, which led them to challenge the male-oriented perspectives they were hearing in class. It raised questions about issues from women's ordination (a movement Johnson joined early on, attending the second Women's Ordination Conference) to the search for a female face of God. Johnson was encouraged in that search by the liberal priests at Catholic University's chapel where she worshipped, who talked about "God our Mother."

Johnson was a pioneer, becoming Catholic University's first woman to earn a PhD in theology, and then, in 1981, the first woman to join the theology faculty. Her field was Christology, which centers on the study of Jesus as the messiah. Now and then, she would "slip in a feminist reading," like "Can a Male Savior Save Women?" from Rosemary Radford Ruether's groundbreaking and still provocative work, Sexism and God Talk (1983). Ruether questioned issues like the use of Christianity to enforce female subjugation and the refusal of the Church—which continues to this day—to acknowledge that a woman can image Christ. "I would have, like, forty [male] seminarians reading this," says Johnson, with a broad smile, "and we would have a wonderful discussion. No one ever objected, not the faculty, not the chairman."

In 1986, Johnson broke another barrier, becoming the first woman eligible for tenure in the theology department at Catholic University. She had the credentials and went through the required "hoops" with aplomb. She got the needed approvals from her department all the way up to the president of the university. Then the application went to the board of trustees—usually a rubber stamp. But because Catholic University is a pontifical university, and 90 percent of the members of the board of trustees are bishops, she reports, "That's where it stopped."

Johnson found herself in a bruising tenure battle, "not with my university," but with Rome. The implicating factors were that she was a woman; that she was not ordained; that she had publicly supported Father Charles Curran, beloved and brilliant professor at Catholic University who was being threatened with firing for his disagreement with the Church's ban on birth control; and that she had written an article—one article—on feminism and Jesus' mother, Mary, that didn't sit well with the Vatican.

In it, Johnson says, she was quoting "some pretty vicious things" being said about Jesus' mother, Mary, by other theologians, like Mary Daly, "like that she's sexless, passive, too obedient, and not a good role model for a feminist." In her article, Johnson reports, she agreed with the criticisms but also attempted to counter them, insisting that "if we went back to Scripture, you wouldn't see her that way." In fact, stresses Johnson, "I was trying to defend Mary! I was on the good side, but they couldn't see that."

The late Cardinal James Hickey, who Johnson recalls had closed down the chapel at Catholic University for being too progressive, was then chancellor. Johnson says that he was the most opposed to her appointment. Despite assurances from the university's canon lawyers that Johnson's being a woman was not a problem (the Gregorian University in Rome already had a tenured woman), Hickey wanted validation from Rome. So Hickey sent all of Johnson's writings—sixteen pieces in scholarly journals—to the Vatican's Cardinal Ratzinger "for vetting." "He wanted Rome to tell him it was okay [to give me tenure]," recalls Johnson. "And of course, they told him it wasn't okay."

The Vatican sent Johnson a list of forty questions—which they called "dubia" or doubtful things—regarding her theological beliefs about Mary, questions like, "You say Mary's too passive. Isn't obedience the greatest virtue?" She answered every one—and waited. Her peers got their letters of congratulations on gaining tenure. Instead, Johnson got a call from the dean.

Hickey wanted to see her. But not just Hickey. On orders from the Vatican, she was to appear before a committee of all the American cardinals to be interrogated about her theology.

Round One: Johnson v. the Cardinals

On a September morning in 1987, the day of her interrogation, Johnson headed for the Army-Navy Building in Washington, D.C. In an indication of the adversarial nature of the proceedings, Hickey insisted the meeting had to be held on "neutral" territory. Johnson wore red, white, and blue, "on purpose." The six cardinals wore "their black things, with the chain," she recalls. All of them sat at a rectangular table, each with a fat three-ring binder before them, holding copies of all of Johnson's writings and related correspondence from Rome. Johnson was the only woman in the room.

The cardinals brought their attendants. Johnson brought her canon lawyers. There was also a court stenographer there, on the advice of Johnson's canon lawyers, "so no one can go out afterwards and misinterpret what I said." And there was an agenda prepared in advance of the meeting, another recommendation of Johnson's lawyers, "to prevent a fishing expedition." Even if they knew she was active in the women's ordination movement, she had not written about it, so they would have no right to ask about those activities.

Not all of the cardinals were antagonistic. At one end of the spectrum was the late Cardinal Joseph Bernardin, then Chairman of the Board of Trustees at Catholic University. "He was just wonderful," says Johnson. "He began by saying, 'I do think the issue of women is one that we need to face in the Church more and more. . . . [When] the time comes, when we actually do face it, Dr. Johnson's work is going to be enormously helpful to us.'"

But at the other extreme was former archbishop of the diocese of Boston, Cardinal Bernard Law, forced to resign in disgrace in 2002 over his protection of priest pedophiles. His opening salvo was hostile in form and content and lowered the bar for their theological discourse. "Does Our Lady pack more of a punch in heaven?" he wanted to know. To Johnson, his question conjured up images of a dysfunctional patriarchal household where Jesus' mother, Mary, would have to intercede with a cranky patriarch who wouldn't deal with the kids himself. In other words, that ordinary mortals best not try to get to this fearsome God directly. It is not a theological vision Johnson shares. "He made me angry at him, personally," she says of Law. "Some of the others didn't. They were there because they were executing Rome's wish . . . but he was personally offensive toward me. . . . He was insulting . . . arrogant . . . ignorant. He was the worst of the interrogators."

Feminism then, as now, was the burning, threatening issue to Church authorities. "Is feminism as important as Scripture and tradition?" the cardinals challenged. Ever the teacher, Johnson instructed them about revelation. "Christ is the source of revelation," she said, "and we learn about Christ through Scripture and tradition. . . . Feminism is a way of interpreting Scripture and tradition from the viewpoint of women, and that has not ever been done before. So in a way, it's a new moment of revelation in the Church."

Later in the session, Law was trying to locate something in his binder to challenge Johnson on but couldn't put his hands on it. He was obviously running out of patience—not helped by the fact that Johnson had just responded to another of his challenges to her, on the singularity of Mary's Assumption, body and soul, into heaven. She pointed him to the doctrinal documents she had decided on her own to bring to the proceedings. She showed that the documents did not define Mary as the only one taken into heaven that way—in other words, that he had gotten the doctrine wrong.

Suddenly, "he slams the three-ring binder shut, like that," she says, demonstrating for me. "And he says, 'Well, you mostly write in Christology. You're not going to do any of that feminist stuff anyway.' " Even then, Johnson knew that was not so. "Something in me just said: '*Oh. You wait and see.*' "

When it was all over, Johnson felt relieved but also "furious, insulted . . . violated in every way." For weeks afterward, she didn't know if she should start looking for another job. She was officially on notice that she was being fired, which was what would happen if she failed to get tenure. Then one day as Johnson was teaching, the heads of her students turned toward the glass window in her classroom door, where the dean was giving her a vigorous thumbs-up. He reported that the board had voted her tenure. "The students gave me a standing ovation in my class, right then and there," reports Johnson, beaming. "Then the next class I came in, there were flowers [and] congratulations all over the board."

Johnson was pleased, of course. But that afternoon, she left class, and the dam broke. "I walked along the Potomac River, just sobbing," she recalls. That is "when my anger really surfaced. I could really face what I had been going through all those months. . . . It was a disrespectful and insulting process. I, who had given my life to the Church, first of all as a vowed religious, and then as an excellent teacher, the best in the department, and as a wonderful, a budding young scholar. They should have applauded me!"

Before that, Johnson admits, she had been "trying to be the good girl. Obviously I wasn't doing everything they wanted. But [I was] still thinking, you know, 'I should be.'" But that experience brought her "eyeball to eyeball with the patriarchal system in the Church and how it works. . . . I saw how corrupt it was from within—not persons, look at Cardinal Bernardin—but the system that would allow that to happen . . . that would not stand for different values." It changed everything for her. She began to think, "Why would I want their approval?"

"I'll tell you when I really began to heal," she says, leaning toward me conspiratorially. It was during another walk along the river. "I said to myself, 'You know what? You know what?'" Johnson lowers her voice and bangs on the table with her fist. "'I'm guilty as charged!'" she says, laughing out loud, tossing her head back. "'*Yes*, I am a woman. And I am interested in feminist theology. And I *do* support Charlie Curran. So, you were right to be suspicious because I am not trustworthy, according to your norms!' I owned myself in that moment. And it's been no turning back ever since."

The year after she won her tenure battle, Johnson left Catholic University. Though she had survived that grilling, the experience had left its mark. One reason she departed was the fear of self-censorship. "How could I go on publishing and thinking and growing as a scholar if I would be afraid that every thought I had was going to be punished?" she says. "I had to go to a place where I would be free to think and develop as a young theologian, even think outside the box, rather than just keep uttering the party line. I had to or I would have died intellectually and spiritually." She found that freedom at Jesuit-run Fordham University in 1991.

The other reason Johnson left Catholic University was Father Charles Curran. "The same year I was fighting for tenure, Charlie Curran was on trial for his positions on birth control, and he got fired," she says. "He lost his battle, and I won mine. I thought: 'How can I stay and give the best energies of my life to a place that has no room for the likes of him?' . . . They were crushing the most creative thinkers."

Johnson went on to write a blockbuster, theologically speaking, *She Who Is*, fueled by the hurt of her undressing. Published in 1992, it is a book whose visions jump off the page, images pour out of her, one after the other, of a new God, a loving, nurturing female God. The book is stunning, original, and healing, and well within the bounds of Catholic teaching. Johnson *discovers* the female God. She doesn't invent her. Or contrive her. She finds her lying in wait, all these centuries, in the pages of everyone's well-worn Bible. In the Old Testament, she finds Wisdom-Sophia marching through time, full of love and self-righteous indignation, compassion and hope, an adoring mother, containing all creation. Johnson explores God's capacity for pain; the power and infinite depth of God's connection to us; the way God loves us, at times like the mother of a suffering child. She describes this female God as "one whose very nature is sheer aliveness . . . the freely overflowing wellspring of the energy of all creatures."

The book's title refers to the burning bush that Moses saw, the God (Exodus 3:14) who identifies as "I am who I am," the God who promises to be Moses' partner in deliverance and liberation. Johnson argues for calling this God not "he who is," the usual translation, but "she who is." Contends Johnson: "[L]inguistically this is possible; theologically it is legitimate; existentially and religiously it is necessary if speech about God is to . . . be a blessing for women." But not just a blessing for women. "If the mystery of God is no longer spoken about exclusively or even primarily in terms of the dominating male," she writes, "a forceful linchpin holding up structures of patriarchal rule is removed."

I am deeply touched by Johnson's insistence on this female imagery for God, but it fails to translate into my prayer life. In my efforts to let go of God the Father, I find nothing there but empty space. I confess this to Johnson, along with my struggles about how to begin. It is similar to the question I hear a frustrated Catholic religious education teacher ask her at a meeting of Catholic directors of religious education in Queens, New York. How, he wants to know, do we teach the female face of God?

She directs all of us to the Bible, which she has mined for female images of the divine. "Jesus called God 'Father,' yes. But he said a lot of other beautiful things about God," she explains. She refers to the parables of the lost sheep and the lost coin (Luke 15:1–10). Why favor one, she asks, God as the male shepherd searching for his lost sheep—which is a fixture in the popular imagination—and not the other, God as the woman looking for her lost coin? Both convey the same message of God's commitment to each one of us, of God's joy in bringing just one lost soul home. She talks about the bakerwoman parable (Matthew 13:33), where God is like a bakerwoman and the kingdom of heaven like her yeast, and about Jesus referring to himself in female imagery, "as a hen gathers her brood under her wings" (Matthew 23:37–39). Then, of course, there's Genesis 1:27: "God created humankind in his image, in the image of God he created them; male and female he created them." Given that, Johnson asks: "Why *can't* God be in the image of a woman?"

But Johnson also recognizes the resistance. We're so used to thinking of God as father that we find the alternatives "almost laughable," she says, and certainly optional. This is not what Johnson thinks. She sees this "godwork," which is what she calls it, as urgent. If we fail to do it, she says, "We're going to forget that these are only images and make them absolutes. We're going to forget that God is God and cannot be contained in any particular image."

I'm trying to find this God, but it's not easy. I don't know who is at the other end of my prayers anymore. What does this female God look like? I can't seem to get the white guy with the long beard out of my brain. It makes me sad that I can't imagine a feminine divine, and curious that when I do, I worry that I'm being equally exclusive as those who imagine only a male God. I want a genderless Spirit, but then I see nothing, no one, and I reach back for the white guy with the beard, disappointing myself. "You go through . . . what John of the Cross called 'the dark night of the soul,'" Johnson says. "You lose your usual moorings. But if you keep walking, she's there. You know what I mean?" Johnson leans forward, head cocked, eyes intent on mine: "She will bring you forth."

Round Two: Johnson v. Cardinal Ratzinger

Johnson's next round with Cardinal Ratzinger and his Congregation for the Doctrine of the Faith began in 1995, after he had issued a declaration that the ban on women's ordination was infallible. She was no longer the neophyte she had been when she walked into the cardinals' den. She was the incoming president of the prestigious Catholic Theological Society of America (CTSA), a 1,400-member organization that had gone from no women members at its founding in 1946 to an estimated one-third of its members today—and one-half of its annual conventioneers and its most active members.

"A lot of people in the society were very upset," recalls Johnson. "The head of the Congregation for the Doctrine of the Faith had never come out before and said, 'I'm declaring this to be infallible.' We were saying, 'You show us where in the history of the Church someone has ever tried to do this before, and declare something infallible, not by the pope.'" Johnson, with the Board, decided that this was an issue to which the society had to respond. She appointed a task force to write a statement addressing the Vatican's declaration, including among her appointees Yale's Sister Margaret Farley, another pioneering feminist theologian who would later head up the society. The board and Johnson set aside funds to arrange for the task force to meet, which the task force did and drew up a statement. Johnson brought that draft statement to the board and the full membership for feedback. The final document went from the task force to the board to the business meeting at CTSA's 1997 convention, where it was approved by a vote ratio of ten to one. The members also agreed to go public. It was a dramatic step for the society. "It puts everybody on the line," says Johnson. "It's a strong thing to do. And we don't do it often."

In 1997—when another woman, Jesuit School of Theology's Sister Mary Ann Donovan, was president—the CTSA issued its statement. "There are serious doubts regarding the nature of the authority of this teaching and its grounds in tradition," it declared, and "serious, widespread disagreement on this question not only among theologians, but also within the larger community of the church. . . . Further study, discussion, and prayer regarding this question by all the members of the church . . . are necessary."

The statement also challenged the congregation's contention that there is a basis in biblical history and unbroken tradition for the refusal to ordain women. It noted that many women were active in the ministry in the early church; that Jesus' naming of the twelve apostles (understood by many biblical scholars to be a symbolic representation of the patriarchs of the twelve tribes of Israel) did not constitute ordination; and that while the Catholic Church has traditionally excluded women from the ordained priesthood, not all traditions are "legitimate," particularly those based on women's "inferiority" and "divinely intended" subordination.

The society "took a lot of flak" for publishing the statement, admits Johnson. The U.S. bishops denounced the paper, declaring that it was "doctrinally and theologically unthinkable, as well as pastorally irresponsible," to reject the restriction of ordination to men. Fiercely defending the Church's exclusion of women from ordination, former Cardinal Bernard Law of Boston decried the society as "a theological wasteland." But Johnson was proud of the work the society's members had done together.

"Even in *Humanae Vitae*, the encyclical upholding the Church's ban on birth control, Paul the Sixth writes in there, 'This is not infallible,'" Johnson concludes. Claiming a teaching to be infallible is "too strong a statement to make . . . unless you're absolutely sure." By declaring the ban on women's ordination to be infallible, the Congregation for the Doctrine of the Faith was "playing dirty." They were trying to force the teaching on the Church. "What we basically said was: 'You can't do that.'"

Today's Challenges to Feminist Theologians

Today, Catholic women theologians are training the Church's future priests and bishops, as well as undergraduate students, graduate students, and many lay ministers. They teach all over the country—in Catholic seminaries (freestanding institutions run by a diocese or religious order),

Protestant or interdenominational seminaries (e.g., Union Theological in New York), and the theology departments of universities (Catholic and non-Catholic).

Not surprisingly, Catholic seminaries—where priests are trained—have become increasingly difficult environments for feminist theologians. Some of those theologians, particularly nuns, have become the targets of institutional purges. One such target was Sister Carmel McEnroy, formerly a tenured theology professor at Saint Meinrad School of Theology in Indiana. She was summarily fired after fourteen years in that job, with less than two weeks notice. The charge: "public dissent" from magisterial teaching on women's ordination, sparked by her signing an open letter, sponsored by the Women's Ordination Conference and published in the *National Catholic Reporter*, calling for continued dialogue. The school, which let her go in 1995, claimed that McEnroy's action made her "seriously deficient in [her] duty." She sued the university for breach of contract but lost. Undeterred, McEnroy went on to publish the widely acclaimed *Guests in Their Own House: The Women of Vatican II*, where she recounts her ordeal.

A few years later, Barbara Fiand, a sister of Notre Dame de Namur, was fired from her teaching position of seventeen years at Mount Saint Mary's of the West Seminary at the Athenaeum of Ohio, reportedly one of the most progressive seminaries in the country. The reasons, she told the Catholic press, were "accusations from unnamed sources that she does not support vocations of future priests to the ordained ministry as practiced in today's Church"—charges Fiand denied. At the time, Fiand was the only remaining female, full professor at the Athenaeum.

In 1995, the Vatican's long investigation of Sister Ivone Gebara, Latin America's leading feminist theologian, finally came to an end. It began over her defense of abortion in some circumstances for the poor women with whom she lives and works in a barrio near Recife, Brazil. Then it moved on to attacks on her theology, based, she reports, on surreptitiously obtained tapes and materials from her workshops that were secretly passed on to the Vatican. Compliments of the Vatican Congregation for the Doctrine of the Faith, Gebara received pages and pages of questions asking her to account for her theology—for example, her nontraditional image of God as universal and cosmic, rather than patriarchal and almighty, and her vision of the Trinity as reflecting "unity and multiplicity," not something outside us, but something that contains us all.

She answered their questions three times before giving up in despair. At that point, the Vatican pressured her congregation (Sisters of Notre Dame—Canonesses of Saint Augustine) to act. The decision was made that if Gebara wanted to stay in her community, she must be sent, like a transgressing child, to be "reeducated" in proper Catholic theology. At the time, Gebara held a doctorate in philosophy and had taught for many years at the Catholic Theological Institute of Recife. She was to be silent, leave her home in Brazil, and move to France to study. Her supporters called it a "forced exile."

That exile lasted one year. But Gebara turned her punishment to her advantage by writing a blistering book, based on her dissertation at Catholic University of Louvain in Belgium, the pontifical university where she was sent to study. *Out of the Depths: Women's Experience of Evil and Salvation* is an ode to the suffering of poor Latin American women at the hands of the Catholic Church and society. The book is rooted in Gebara's belief that the Resurrection, not the glorification of suffering and the cross, is the Church's most central teaching. It is an unmistakable rallying cry to Latin American women to turn their Christianity from a holy reason to bear their crosses and suffer in silence to a holy motivation, through the life of Jesus, to rise up and change their lives. The book also reflects Gebara's journey through her own trials at the hands of the Church to her own salvation. In a noteworthy irony, Gebara's dissertation was signed by the pope. "I wonder if he even read it," she says to me, shaking her head. Undaunted, she is back in Recife, teaching and writing, delving deeply into ecofeminist theology—which links ecology, spirituality, and feminism—and serving on the board of *Catolicas por el Derecho a Decidir*, Latin America's version of Catholics for a Free Choice.

Seminary professors like Sister Katarina Schuth, who studies the state of Catholic seminary education and teaches at Minnesota's Saint Paul Seminary School of Divinity, University of Saint Thomas, reports being quite satisfied with her position (though not with the conservative direction the Church is taking) and believes that many other female seminary professors are satisfied, too, particularly the liberal ones at the more liberal seminaries and the conservative ones at conservative seminaries. But many reformers do not believe that is the general sentiment among female Catholic seminary professors—particularly those who are outspoken fighters, which Schuth admits she is not. In a statement of protest about the firings of female seminary professors, the National Coalition of American Nuns wrote: "No matter how impressive their academic credentials, their teaching records, their publishing abilities, or their history of fidelity to the Church, there seems to be no room for women professors on seminary faculties. They have become a threat to the clericalism . . . which too many seminarians today desire, profess, and exemplify."

While Elizabeth Johnson thrives in her position in Fordham University's theology department, the experience of Catholic feminist theologians teaching in the theology departments at Catholic universities—as opposed to Catholic seminaries—can be "a mixed bag," according to Susan A. Ross, tenured professor of theology at Loyola University in Chicago. She is deeply appreciative of the "basic sympathy with a faith orientation that you don't have to apologize for" that she finds at Loyola. It is a spirit that can be lacking, she observes, at non-Catholic institutions, where, she says, "You can detect . . . a bias against religion on the part of a lot of academics . . . particularly a bias against Catholicism." That bias, she goes on to explain, "is in some ways deserved because of the way the Vatican treats intellectual discourse, but it's also a kind of snobbery."

By taking theology seriously, Loyola has provided Ross with "the opportunity to ask all the hard questions," she says. She educates her students on both sides of the most controversial issues, from ordination to abortion to birth control. Her aim is have her students be "well-informed . . . so they know that, even while they may disagree with, say, the Catholic Church's position on the ordination of women—which I totally disagree with—at least they know that it's not because, quote: 'The Catholic Church hates women.'" She reports that she feels "no sense of limitation [in terms of what she can teach] . . . absolutely none at all." Because she has reached the level of full professor and received various accolades and a coveted teaching award, she feels supported by the university and that "my accomplishments, in many ways, have been recognized.'"

At the same time, Ross has crashed into the stained-glass ceiling. When it came time to find a new chair for the theology department, she was elected by a majority of the members of her department to fill that position—the first woman ever elected. However, the administration refused the department's recommendation, instead appointing a Jesuit from the outside. One observer who chose to remain anonymous reported that names like Elizabeth Johnson's came up for consideration, but she was deemed to be "not Catholic enough."

Ross suspects that her feminism was implicated in her loss of the chairmanship, as well as a critique she had written in the mid-1990s of the Jesuits' statement on women and the Church in civil society for a Jesuit magazine, which she only wrote after being requested to do so. "I know that I've antagonized some people who think of me . . . as a bitter, angry feminist," says this wife of a former Jesuit. "I mean, if they think *I'm* a bitter, angry feminist, they haven't met one."

Many feminist theologians no longer teach at Catholic institutions because of a climate hostile to feminism and other creative thinking. Another potential discouragement has been the Vatican requirement that theology professors obtain a "Mandatum"—that is, the local bishop's written permission to teach at a particular university and to do so only according to "authentic Catholic doctrine"—which actually is very unevenly enforced. One result of this discouragement of feminist theologians can be a great loss of intellectual energy and vigor at these Catholic institutions.

By contrast, Catholic women theologians who teach at Protestant and nondenominational theology schools have an unprecedented opportunity to push the frontiers of Christian theology well beyond its traditional boundaries. They are out of reach of the institutional Church altogether, free to explore theology, without the threat of backlash from administrators or the hierarchy. Rosemary Radford Ruether, for example, began her career in Howard University's theology department, where she developed a race as well as a gender analysis of oppression within Christianity. She then moved to Garrett-Evangelical Theological Seminary in Evanston, Illinois, from which she retired in 2002; she is now at the Graduate Theological Union at Berkeley, California. Says Ruether: "I am a Catholic woman theologian, but I am not harassed by the hierarchy. They don't bother me because they don't have any power over me."

But Ruether has been effectively disowned by some in the institutional church, at times banned from speaking in Church venues and regularly demonized by the Catholic right wing. But she has legions of supporters among Catholics and non-Catholics alike. On the occasion of Ruether's retirement, Loyola's Susan Ross wrote, "The voices of women like Rosemary Ruether are as much a part of the Catholic and Christian tradition as are the voices of the hierarchy, and Rosemary's work has inspired me and many others to speak and to write, to teach and mentor, in ways that we hope will transform the Church in years to come."

In addition to teaching theology to clerics, as well as lay Catholics and non-Catholics, many of today's Catholic feminist theologians have become major voices on the world theological stage. They are interviewed by print, television, and radio journalists. They speak before numerous and varied groups worldwide. Their books appeal to diverse audiences, not just academics. "Look at Elizabeth Johnson, Rosemary Radford Ruether, Lisa Sowle Cahill, Margaret Farley, Shawn Copeland, [the first African-American woman president of the Catholic Theological Society of America]," says Sister Sandra Schneiders, a provocative feminist theologian who teaches at a more liberal Catholic seminary, the Jesuit School of Theology at Berkeley. "Those are big names. . . . They're not in-house trained to train in-house. They're publicly functioning theologians." And that has an impact well beyond academia.

The Bridge from Academia to the Pews

On a wet, cold night in March 2004, a stream of people climb the steep stairs to Saint Francis Xavier Church in Manhattan for a Voice of the Faithful meeting. Hundreds of people settle into the pews. It's a much larger crowd than those that usually come to the meetings, where they struggle to try to figure out how to make contact with a remote and hostile hierarchy, led by Cardinal Edward Egan, who might as well, in terms of VOTF, be living on Mars. The reason they are all gathered here tonight: Elizabeth Johnson has come to speak.

In recent months, Johnson's new book has come barreling off the front pages of major Catholic magazines. "Mary, Mary, Quite Contrary" blared the headline for U.S. Catholic's cover story, accompanied by the photo of a woman in Middle Eastern garb, back to the reader, raising a clenched fist. With that book, Johnson returned to the subject that got her into trouble with the Church authorities in the first place: Jesus' mother, Mary. Called Truly Our Sister, the book is an appeal for the humanity of Mary, who Johnson locates, as she did with the rest of us in her earlier book, in the communion of saints, as "truly sister to our strivings."

Johnson resists the impulse, apparent around the globe, to turn Mary into God, advising instead that we let God have her own maternal face, while keeping Mary for ourselves. She wants to end the separation of Mary from humanity, particularly from all other women. She sees this as achieved by the Church's unending emphasis on the Virgin Birth, the Immaculate Conception, and the Assumption of Mary into heaven. Writes Johnson: "The doctrinal-mythologizing approach, which governs Catholic tradition, separates Mary from all other women. . . . [By] making Mary here the exception rather than the type, these doctrines . . . subtly disparage women's sexuality, holiness, and independence."

As she did all those years ago before the cardinals tried to clip her wings, Johnson presents a proud and valiant Mary, a fighter for the poor and weak, for liberation from the bounds of oppression. She does this by focusing on Mary's Magnificat (Luke 1: 46–55), the revolutionary song of praise that Mary proclaims after she agrees to the visiting angel that she will bear Jesus. In the Magnificat—which begins "My soul magnifies the Lord" and says, "Surely, from now on all generations will call me blessed"—vanished is the submissive Mary. What emerges is a vocal, proud, and vigorous young woman. She recognizes immediately that what God has done for her—"he has looked with favor on the lowliness of his servant"—is what she, God, and her child can do for an ailing world. "What begins as praise for divine loving-kindness toward a marginalized and oppressed woman grows in amplitude to include all the poor of the world," writes Johnson.

Despite the power of Mary's Magnificat—which is the most any woman gets to say in the Bible, notes Johnson—many Catholics never hear those words. In the course of years of research to prepare for teaching Marian theology, Johnson reports that she "never saw these great theologians of the past who did all this Mariology make use of the Magnificat. . . . They made their choice of texts, and it was always Mary conceiving Jesus, but not what followed, her proclamation. . . . And you can see why, when you see how strong it is." The Church lectionary attests to this: The Magnificat is buried; it is never assigned to be read at a Sunday Mass.

At the end of her presentation to her VOTF audience in Manhattan, people have a lot of questions. There is deep pain in this group over the sex abuse crisis and a longing to act. The last questioner asks, "Given the crisis in the Church today—the abuse of children by priests, the cover-up of those crimes by our own bishops—if Mary were here, what do you think she would say?"

At first, Johnson seems surprised by the query. She stands silently, thinking. Finally, she replies. "The Magnificat," she says, as if it is self-evident. "He has brought down the powerful from their thrones," she says, "and lifted up the lowly."

As they study the realm of the sacred and the meaning of God, some Catholic feminist theologians maintain a deep fidelity to Catholic dogma. Others are more fearless, questioning the basic tenets of the faith, from Jesus' singular divinity to the widely challenged Virgin Birth, from papal infallibility to the place of Catholicism among the world's religions. Instead of a commitment to blindly maintaining a particular system of theological thought, they are embarked on a thorough and unbridled search for truth.

What the ultimate impact of this work will be remains to be seen, but it will surely be dramatic. While pioneering feminist theologian Mary Daly has rejected institutional religion entirely, her words from a much earlier time remain prescient today: "As the women's movement begins to have its effect on the fabric of society, transforming it from patriarchy into something that never existed before . . . it can become the greatest single challenge to the major religions of the world, Western and Eastern. Beliefs and values that have held sway for thousands of years will be questioned as never before. This revolution may well be also the greatest single hope for survival of spiritual consciousness on this planet."

In his book *A People Adrift*, religion writer Peter Steinfels makes a similar observation: "Over the long run, nothing in Catholic Christianity, like nothing in other forms of Christianity—or in Judaism, Islam, Buddhism, and even Hinduism—will remain untouched by the passage from a patriarchal era to one of female equality. . . . [N]o one should be confident about predicting which elements in the great religions will be radically revised, which will be reconfigured, which will remain relatively intact."

For now, the work of feminist theologians is challenging and reshaping Church teachings. At the same time, it is having an impact on Catholic ministry, particularly women's place in that ministry. While tearing down the flimsy arguments that bar women's full participation, the theologians, in partnership with the activists, are illuminating the real roles women assumed in the early Christian Church. By doing so, they are providing an irrefutable foundation for change.

11

✠

CATHOLIC WOMEN
RAISE THE DEAD

Key to the Vatican's opposition to equality for women in positions of authority—from deacons to priests to bishops to cardinals to the pope—is the historical obliteration of the central ministerial roles played by women in the ancient Church. One of the women who has done a spirited job of raising those women from the dead is Sister Christine Schenk.

A preacher by nature, given to far-off gazes and pontificating in a slow midwestern drawl, Schenk wasn't always a nun. In fact, until she turned forty, she was a woman of the world. She owned her own home and had a good job as a midwife in a program she helped to start at Case Western Reserve University Hospital. And she made "big bucks, for Cleveland." She says, "I was rapidly becoming independently wealthy." But something was missing. Schenk joined a religious congregation and, in 1988, became a nun. At about that time, she also began to pursue a lifelong dream of studying theology. That is what, she says, "radicalized" her.

Using the groundbreaking biblical research of many feminist theologians, Schenk helped to launch a drive to restore women to their rightful place as priestly ministers, apostles, deacons, disciples, preachers, and prophets in the early Christian Church. Before Dan Brown—author of the mega-bestseller *The DaVinci Code*—put Mary Magdalene front and center in the life of Jesus Christ, Chris Schenk was working to drag her out of the bordello and onto the altar. As cofounder and head of FutureChurch, Schenk helped to birth an international movement to reeducate the faithful about the real biblical Mary of Magdala and about a growing array of courageous, tough, and influential Christian women.

Celebrating Women Witnesses

"Celebrating Women Witnesses," a joint Call to Action/FutureChurch program, promotes Christian women of faith, both ancient and contemporary, by providing accessible information about them to Catholics in parishes worldwide, based on the work of eminent theologians. The goal "is to educate about the radical inclusivity of the historic Jesus," says Schenk, to bring to a wide public the biblical scholarship showing that Jesus indeed had women in his closest discipleship.

Among those most eminent theologians is Elisabeth Schüssler Fiorenza. This woman who stares out at the world through bright blue eyes under a cap of white hair, began her career in Germany in the 1960s, among the first woman to earn the equivalent of a master's degree in divinity, followed by a licentiate degree in practical theology and a doctorate in New Testament studies. She came to the United States with her theologically trained husband to teach with him at Notre Dame and soon exploded onto the theological scene with her writings. She would become the first female scholar to serve as president of the Society of Biblical Literature, the nation's oldest and largest biblical society. Today, Schüssler Fiorenza (never call her "Dr. Fiorenza"—"That's my husband's name," she'll scold) sits in a real ivory tower, her third-floor office at Harvard Divinity School, where she teaches, which has wide windows overlooking a sloped lawn. In a testament to her fear for women's progress, a sardonic little cartoon hangs on her door. Under a banner that says "Post-feminism," the cartoon reads: "Keep your bra. Burn your brain."

Schüssler Fiorenza set a new standard for biblical research with her groundbreaking work of biblical reconstruction called *In Memory of Her*. She took the title for her book, published in 1983, from the passion account of Mark's Gospel (14:9), "The Anointing at Bethany." In it, Jesus is anointed by a woman who carries "an alabaster jar of very costly ointment," which she pours on his head. This sends the disciples into a rage, insisting the ointment should be sold and the money given to the poor. But Jesus rises to her defense. "She has done what she could; she has anointed my body beforehand for its burial," he said. "Truly I tell you, whenever the good news is proclaimed in the whole world, what she has done will be told in memory of her."

Schüssler Fiorenza interprets this anointing of Jesus' head as a prophetic act, akin to the Old Testament prophets who anointed kings. That action "must have been understood immediately as the prophetic recognition of Jesus, the Anointed, the Messiah, the Christ," she writes, as well as one whose "messiahship means suffering and death." But Jesus' instruction regarding the woman who dared anoint him was soon ignored. While the stories of Judas, who betrays Jesus, and Peter, who denies Jesus, both in the same Gospel of Mark, "are engraved in the memory of Christians," writes Schüssler Fiorenza, "the story of the woman is virtually forgotten. . . . Even her name is lost to us."

In the intervening years, Catholic and non-Catholic feminist theologians have turned their analytic eyes to reconstructing the lives of women in the early Church. They have used traditional historical resources, including biblical texts, but also have delved into newly uncovered ancient documents. Those documents illustrate women's powerful roles in the early Church, documents that were so threatening to early Church leaders that they were literally buried for more than a millennium, until an Arab peasant looking for fertilizer accidentally discovered them in 1945. Called the Nag Hammadi Library (for the town in upper Egypt where they were found), one set of these documents includes the primary Scriptures of the Gnostics, a early and vocal Christian sect whose dangerous ideas included beliefs in the female divine and in Mary Magdalene's primacy as a consort to Jesus and an early Church leader whose power rivaled the apostle Peter's. So threatening were the Gnostics that the early Church fathers declared them heretics, suppressed their writings, and kept all trace of their ideas out of the official biblical canons.

Elaine Pagels was the first feminist theologian to popularize these extraordinary finds in her bestseller *The Gnostic Gospels*; most recently, she brought another of the buried Scriptures to light with her book *Beyond Belief: The Secret Gospel of Thomas*. Karen King is another non-Catholic scholar working in this field today, author most recently of *The Gospel of Mary of Magdala*. Bernadette Brooten is one of the major Catholic scholars. Her research led her to determine that the male apostle who had appeared as "Junius" in earlier versions of the Bible was actually a woman, Junia—a discovery made official with the New Standard Revised version of the Bible in 1989. Despite the Church's insistence that all of the apostles were men, Brooten's discovery insures that in the Bible today, in black and white, at least one female apostle lives.

More Buried Treasures

While rewriting Church history is one method of restoring women leaders to the ancient and evolving Church, another method is archaeology—that is, the study of ancient artifacts and the archaeological dig. Dorothy Irvin holds a pontifical doctorate in Catholic theology from the University of Tuebingen in Germany, with a specialization in the Bible, ancient Near Eastern studies, and archaeology. A tiny, bold, and earnest woman who stands close to you when she talks, she began her field experience in archaeology in 1966 and later joined the staff of the Madaba Plains Project in Jordan, where she has been active for seventeen years. Her interest in early Christian archaeology and women's history led her to create her first calendar, in 2003, *The Archaeology of Women's Traditional Ministries in the Church*. It features photographs of frescos, mosaics, and tomb inscriptions showing women deacons, priests, and bishops from 100 to 820 A.D. In her 2004 calendar, she continued with those photos, moving up to 1500 A.D., and in 2005 added a map of Europe and the Mediterranean locating the sites of the artifacts.

The calendar imagery is stunning, the messages incontrovertible. A fresco in a small underground chapel, in the Catacomb of Priscilla in Rome, pictures a group of seven women, seated along a table, in what is obviously a Eucharistic celebration. At either side of them are seven bread baskets—an early Church symbol for the Eucharist. Four of the women have their hands outstretched over a plate and cup, "in what is still familiar to us today as the gesture of consecration during the liturgy of the Eucharist," writes Irvin. The fresco is dated around 100 A.D.

Irvin presents the tombstone of Sofia the deacon, whose fourth-century inscription reads: "Here lies the minister and bride of Christ, Sofia the deacon, the second Phoebe," a reference to the deacon Phoebe, to whom Paul refers in Romans (16:1–2). Though both women clearly were deacons, Catholic women are forbidden today to hold that office. In a side chapel of the Church of Saint Praxedis, "scarcely a ten-minute walk from the main railroad station in Rome," writes Irvin, is a "beautiful group portrait in mosaic of four women ministers." One of them is Theodora. Above her head is the word *episcopa*—meaning woman bishop. The year is 820 A.D. And in Naples, in the Catacomb of San Gennaro, is the tomb fresco of Vitalia. Wearing a vivid red chasuble, her arms outstretched, she stands at what appears to be an altar. In addition to bread and wine, the fresco shows two open books with the names of Joannes, Marcus, Matteus, and Lucas written across the top, obviously the Gospel texts. The symbols, writes Irvin, "indicate that she celebrates both the liturgy of the word and the liturgy of the Eucharist, that is, that she was an ordained priest."

"The pictures shown are not an artist's imagination or a modern reconstruction of what might or should have been," Irvin reminds us. "They are the early Christian communities' records, inscribed or painted, of what was actually happening. . . . The women shown or named in these sources were not pagans; they were the earliest Christians. They were not heretics or schismatics but were in conscious succession from the apostolic founders of ordained ministries."

Not only are early female Church leaders buried in tombs and catacombs but also in the Catholic Church's lectionary. That is the collection of biblical passages authorized by the Church hierarchy to be read at Mass, according to the liturgical calendar; they rotate over a three-year period. The current lectionary was developed in 1969 by the Sacred Congregation for Divine Worship in response to a mandate from the Second Vatican Council to bring more of the Bible to the people. While "the widely held assumption has been that the lectionary faithfully represents the essence of the Bible, with the omission of only a few troubling or gory passages," as theologian Ruth Fox has noted, in fact, that's not the case. What has been left out entirely or treated as optional are vivid stories of women's central place in the early Christian Church. "Women's [Old Testament] books, women's experiences, and women's accomplishments have been largely overlooked," writes Fox.

Catholics are very unlikely to hear Jesus' instruction to keep alive the memory of the woman who anointed him at Bethany, the woman Schüssler-Fiorenza resurrected. As Fox reports, that passage is omitted from Matthew's Palm Sunday Passion reading and optional in Mark's Palm Sunday Passion reading. John's account is read only on Easter Monday. Conversely, on one Sunday every three years, Catholics will hear Luke's account of the anointing of Jesus by the woman who is a penitent sinner and washes Jesus' feet with her hair.

Catholics are also extremely unlikely to hear on any Sunday about Jesus' healing of the hemorrhaging woman, which Fox maintains is a very significant omission because it illustrates Jesus' disregard for the misogynistic practices of his time, particularly the uncleanliness of a bleeding woman and the taboo against touching her. That reading is optional and only comes up for consideration every three years.

And Catholics never hear Romans 16:1–19. Given the Church's stubborn exclusion of women from ordination to the deaconate and the priesthood, it's easy to see why. There, Paul lauds a long list of women leaders in the early Church, most notably, "our sister Phoebe, a deacon of the church at Cenchreae," and Junia, who is "prominent among the apostles," who was "in Christ before I was." While Catholics will hear about Junia once every three years at a Saturday mass, they will not hear about Phoebe. Furthermore, while the *New Revised Standard Version of the Bible, Catholic Edition*, still refers to her as a deacon, in the *New American Bible*, the only Bible currently approved by the Catholic Church, Phoebe the deacon has vanished; she is now a minister.

During the Easter season, on each Sunday, readings from Acts focus on the activities of Peter, Paul, Barnabas, and Stephen. But on no Sunday will Catholics hear about the women in Acts—Tabitha, Lydia, and Prisca—who rate only weekday readings. "In fact," writes Christine Schenk, "most Catholics are completely unaware there even were women leaders in early churches."

As for Mary Magdalene, all four Gospels show her as the primary witness to Jesus' Resurrection and the one Jesus chose to deliver the Good News to the others. Yet vanished completely from Easter Sunday Mass is the gorgeous encounter between Mary of Magdala and Jesus (John 20:11–19) after his death. Mary stands weeping outside the tomb, bereft that "they have taken away my Lord." Then she turns around to see the figure of a man, Jesus, who she mistakes for the gardener. Jesus calls her by name, and then she realizes who he is. He tells her "Do not hold on to me," a remark evoking the closeness between the two, and "go to my brothers and say to them: I am ascending to my Father and your Father, to my God and your God." Instead, the Easter Sunday Gospel that is read from John focuses on the rush to the tomb of two male disciples who never even see Jesus. The reading ends abruptly just at the moment that Mary's encounter with Jesus begins. "Inexplicably," Schenk writes, "this important account of Jesus' appearance to Mary in the garden does not rate an appearance on any Sunday of the Easter season but is assigned instead to a weekday, Easter Tuesday." Adding insult to injury, Jesus' appearance to Thomas—who had doubted him—is read on the second Sunday after Easter every single year.

Hands Across Time

While Catholic feminist theologians are uncovering the lives of the Church foremothers, groups like FutureChurch and Call to Action are bringing those lives to the people. They encourage Catholic women to develop and lead new liturgies using a host of inventive materials to celebrate the roles of women leaders in the Christian Church—from the earliest centuries to the present. A major accomplishment of this movement has been to put the feast of Mary of Magdala—July 22—on the liturgical map. Today thousands of people attend hundreds of celebrations worldwide on July 22, which also provide the opportunity to resurrect and honor the lives of more contemporary Catholic women leaders. At those celebrations, Catholic women from the ancient Church come to life in the costumed bodies of contemporary Catholic women, who have a deep commitment to the Church and a stake in owning their place in it.

On a wickedly stormy summer night in Detroit, 2002, some 150 women and men file into the Christ the King Church for a celebration on the feast of Mary of Magdala. The light oak pews are arranged in a half circle facing the altar, where a podium has been set up. The room is ringed with windows, amplifying the sound of the rolling thunder and crashing rain outside.

Marge Orlando, a spiritual director trained in pastoral ministry, mother of eight sons, and president of the Michigan chapter of Call to Action, opens the ceremony with an energetic greeting. She wears a black dress, a white pearl necklace, white pearl stud earrings, and a purple ribbon, the symbol tonight for women's ordination. She begins by thanking pastor Victor Clore for making the church space available—no small gift, since Call to Action's members are frequently barred from Catholic venues.

Orlando introduces Eileen Burns, who developed the liturgy for the night's event. She, in turn, introduces Therese Terns. A junior high-school teacher and theatrical director, Terns is playing Magdalene, the night's MC. Terns is tall and thin, with big, brown flashing eyes and short brown hair. She's wearing black pants, a shirt, and a bright red stole tossed over her shoulder. "Were you there?" she asks, eyes narrowing, swaggering a step or two. "I was. In the eyes of some, we were saints. In the eyes of others, we lived scandalous lives."

With that, she introduces Dorothy Day. The acclaimed author, tireless advocate for the poor, champion of nonviolence, and renowned founder of the worldwide Catholic Worker Movement, whose own personal life ran the gamut from sexual carousing to abortion to devoted motherhood, comes alive in the person of a woman who volunteers at Day House in Detroit, scarf on her head, checkered apron around her waist, glasses dangling. Next comes Sister Thea Bauman, an exuberant African-American woman who lived in 1930s America and brought black sacred song and dance to Catholics nationwide, including to a conference of the U.S. bishops. She is played by an equally exuberant soup kitchen director with a booming voice. The Mexican-born Sor Juana Ines de la Cruz—a revered seventeenth-century woman of letters and vocal advocate for women's intellectual rights in the Church, whose library was burned and voice silenced by the Catholic hierarchy—is played by a young El Salvadoran woman. She arrives at the podium in full nun's habit with a deadly serious expression on her face.

Through it all, Magdalene lurks. When it's her turn to speak, she admits that she's wearing red on purpose, to rag on all those people who insist she was a prostitute. She says she came from an area where people were pretty well heeled, so there was no reason to sell sexual favors. She says that Jesus taught her how to give, not just money, but love. "Jesus could look into my eyes, and I would feel I was the only one that he loved," she says. "But I knew that love was so big and so strong that it couldn't just be held by me." Her response, perhaps, to the contention that Magdalene and Jesus were wed. She shares the laments of women through the ages, saying, "You know, Jesus could be very moody. You had to leave him alone because he had *things on his mind*." And she complains that he was always with the guys, who didn't much like having a woman around.

She confesses to the devastation and the powerlessness that all the women felt at the arrest and crucifixion of Jesus and talks about her encounter with Jesus at the tomb, when she heard a voice call, "Mary." "I never loved my name more than at that moment," she says. "I think maybe that's why we're here tonight. It's not about power. It's simply about call. . . . Each of us is called." Terns is sharing the reformers' vision of a priesthood of all believers, from which no one is excluded—certainly not women—and where everyone has a place.

The service ends with a sermon by another vibrant spirit, Karon Van Antwerp, thirty-one-year-old associate pastor at St. Mary's University Parish on the campus of Central Michigan University and a leader of Call to Action's Next Generation Program. Her delivery is Phil Donahue style as she glides from one end of the altar to the other. She says that when she asked her dad if he was coming tonight, he said no. She assumed he had other plans. But when she asked what he would be doing, he told her "nothing." He'd decided to pass on the event because "it's a woman's thing." Van Antwerp flipped. "I said:'I go to church every week believing some man's life might have something to do with me!'"

She says that this event is not just a woman's thing but is about "sharing our faith, something we don't do often enough. . . . Throughout the country today, men and women are gathering to share the story of Mary Magdalene and others who have worked to build the kingdom. We're gathering to remember their wisdom, to see the way God touched their lives and continues to in our world today. We're gathering to claim our places as apostles and friends of God."

She notes that her dad is in the audience. It turns out he's a former Catholic priest.

The Independents

Some Catholic women committed to bringing women's history to other Catholic women have made it part of their life's work and do so independently. By far the best known is Edwina Gately. A writer, poet, and riveting speaker trained in Catholic theology, she's been described as a "one-woman Church." In person, Gately is a force to be reckoned with, with a bold voice, a dynamic manner, and a dramatic message. She is slim, with bobbed brown hair. An English-born expatriate, Gately's manner of speaking is fast and furious. Her subject when I hear her speak at a 2001 Call to Action workshop is "The Beguines."

"At the time," which was the twelfth century, explains Gately, "there were too many women wanting to be nuns, so the men closed off applications." In response, she says, "These women did what women do best: They got together." They thought that anyone could live an apostolic life, devoted to God, but that they didn't have to be under the Church's thumb. So they formed women's communities entirely independent of the Church. They took vows of poverty but not necessarily chastity and spent their days helping the poor and caring for the sick. Many also preached and wrote vociferously, "in the vernacular," which at the time really put them over the line, says

Gately. "It was a departure from everything in Latin, which nobody read." They had their own money, which they used to create their own residences and their own churches. "They challenged corrupt priests and known sex abusers by refusing to receive Communion from them," reports Gately, thereby making "a large dent in the Church pocketbook because the laity gave economic support to the Beguines instead."

Their numbers rose to the thousands, their influence grew, and the Church became increasingly threatened. In the 1300s, the Church officially declared the Beguines heretics. It forbade the Beguine way of life, deemed them mad, confiscated their property, and destroyed their books. One of them, mystic Marguerite Porete, a bold writer and public speaker, was brought before the Inquisition, imprisoned, and eventually burned at the stake.

Gately is the founder of an independent lay missionary movement herself. She relates to the Beguines. When she visited western Kentucky recently as an invited guest at a religious retreat, she tells us, she was declared a "heretical speaker" by a conservative local Catholic paper. The article listed among her sins wearing a priest's stole; writing a pro-women's ordination article; and characterizing God as "black, white, brown, yellow, gay, and straight." Pickets came out in force, Gately reports, "which I guess they do whenever a heretic comes to town." The same newspaper advised protestors to tape-record her presentation "for evidence."

As for the Beguines, says Gately, "after one hundred years of dynamic feminist spirituality," they had "their contribution obliterated." "Ladies," she says, her voice dropping conspiratorially, "we've been here before."

I'm delighted when I learn that such charismatic female leaders as Gately are a great comfort to young Catholic women, like thirty-one-year-old Katie Bergin. With a master's in divinity—a degree that qualifies her for ordination—from Union Theological Seminary in New York, Bergin is a chaplain at a Catholic girls' high school in Manhattan. Such young women have an even harder struggle than their elders did finding a place in this Church. Exquisitely sensitive to discrimination, they see the Church's systematic defense of women's exclusion from leadership as completely anachronistic in their world; as such, it is a source of great pain to them.

Edwina Gately is one of Bergin's role models. Gately's name comes up when I ask Bergin about her faith life. She says, "Edwina Gately talks about staying 'juicy,' and I like that." To Bergin, that means being "alive, full of ideas, in tune with God and other people." It also means owning a place in the Church, regardless of the roadblocks. Gately calls herself an independent Catholic minister. "Gately tells us, 'Hey, if they don't give you a title, take one,'" reports Bergin. While Bergin is a chaplain right now, she's not sure what she'll be next, so she is clearly impassioned by that advice.

Bergin, too, brings women leaders from the ancient Church into her work whenever she can. When asked to preach before a large crowd at the high school's alumni mass, she took as her subject Mary Magdalene.

In defiance of all the evidence that women in the early Christian Church were apostles, ministers, and deacons, that they served Mass and were called bishop, and that Jesus accepted them equally and wholeheartedly into the ranks of his disciples, the Church hierarchy persists in putting put forth hollow theological arguments for women's ostracism from the Church's highest rungs of sacramental authority and declaring that the exclusion has been ordained by God. At the same time, the Church hierarchy, besieged by a crushing priest shortage, has had to reach out to Catholic women for help in the ministry, help on which the very survival of Catholic ministry depends.

Emboldened by knowledge of their own history in the Church, Catholic women have responded to that call with creativity and enthusiasm. Despite the hierarchy's best efforts to circumscribe their input and control their influence, these women have begun to herald change, modeling a collaborative priesthood of all believers and bringing women once again into the Church's central sacred space.

12

✠

THE NEW FACE OF CATHOLIC MINISTRY

Alexandra Guliano has always wanted to be a priest. "I have felt a call to ministry from childhood," she wrote. "I can only describe it as the most powerful movement, both interior and exterior, that I have ever experienced. . . . I felt it even before I had words to describe it. It was as familiar as breathing and as grounding as the earth beneath my feet."

Like many women who feel they have a call to the priesthood, Guliano was "emphatically told that how I experienced God's call in my life was an illusion, or that I was crazy." Despite this, she wrote in her dissertation study of lay leaders in priestless parishes, "the call persisted, and while I struggle with discerning the specific way to respond to the call, I have ceased to doubt its authenticity."

As with the vast majority of Catholic women ministers, Guliano, age fifty-one, had to chart her own path. She earned a master's in systematic theology from the University of Notre Dame and a doctorate of ministry from San Francisco Theological Seminary, part of the Graduate Theological Union in Berkeley. She began as a lay minister, the director of worship, at a large suburban parish in the diocese of Green Bay, Wisconsin. When the

Church began to allow women to assume administrative pastoral roles, she took a position as a pastoral associate in the diocese of Green Bay and then in the Archdiocese of Seattle, conducting prayer services, chairing meetings, training leaders, and being on-call for pastoral emergencies. She also spent a year ministering as an on-call chaplain for a medical center, obtaining certification for that work as well.

Then, when the Church began to open the position of parish director to laypeople, Guliano applied. Soon after, to her great satisfaction, she was appointed parish director of Sacred Heart of Jesus Parish in the Milwaukee diocese by Bishop Rembert Weakland, where she served from 1994 to 2001. After that term, she faced a time of anguish. She did not know what her next opportunity would be to answer her call in a Church where these appointments are seen as temporary, and there is no job security. Finally, she was given a second parish, Saint Therese's, on the outskirts of Milwaukee.

Guliano and I plan to meet when I visit Milwaukee for the 2003 national Call to Action conference. The week before I am to go, my beloved Aunt Anne, my godmother, died suddenly. On a dreary rainy Wednesday, I climbed the stairs to Saint Lucy's Church in Scranton, Pennsylvania, for her funeral Mass, and a few hours later, buried her. When I traveled to Milwaukee the next day, I carried with me a terrible sadness. That evening, Guliano picked me up and drove us to Saint Therese's for a special service. When I got there I learned it was called "An Evening to Remember." It was to be a memorial service for loved ones lost.

One Woman's Church

Saint Therese's is a modern church, built in the 1950s, a boxlike structure from the outside. It is the spiritual home to some 650 families. It is clearly the pride and joy of Guliano, a short, stocky woman with short brown hair and a reserved though welcoming manner. Asked about her personal life, she says she is single and in a significant relationship. Guliano disappears into her office to get ready for the service. I am ushered inside the church.

It has an A-shaped interior that dives dramatically skyward. To the left is the choir loft, full of women. At the front is a long altar. A brilliant coral cloth covers the altar table, scattered on top with autumn leaves. The pulpit is a rich wood that looks like oak. A large cornucopia on the floor brims with yellow and green squash, corn husks, and acorns, and a nearby basket overflows with fall flowers. A side table holds rows of unlit white candles. Some two hundred people sit in four banks of wooden pews in a semicircle, speaking softly, waiting.

Into the church comes Guliano at precisely 7:00 P.M., wearing a crisp white alb and a cincture around her waist. She chats briefly with a few parishioners, then takes her place on the altar. The service begins. It takes little time to see that this service is not about death but about what we can learn from death about life. A parishioner comes up to read Scripture, beginning with Ecclesiastes 3:19, "to everything, there is a season."

"Whether your loved one died suddenly, without warning, or through a long and painful illness, each is embraced by God and loved by us and being given a prayerful send-off," says Guliano in her sermon. Quoting theologian Karl Rahner, she tells us that "though invisible to us, our dead are not absent. . . . They are living near us, transfigured . . . into light, into power, into love." Borrowing from an anonymous poem, she advises us that "memory is a powerful thing . . . a form of immortality. . . . Those we remember never die."

My throat tightens and tears rise, but I feel comforted in my grief. After the sermon, Guliano calls the name of someone lost, while an image of that person appears on a screen behind her. Loved ones walk in silence from a pew to the table to light a candle. Up comes a nun in an abbreviated habit; a man with white hair, dragging his right foot; a cool young man with a stylish buzz cut and a trendy raincoat; two young women maybe in their thirties; a middle-aged Hispanic man; a mother and an adolescent boy; and a woman with a small child. I wish that my aunt's face could be up there on that screen, that I could light a candle.

Afterward, parishioners serve cider and cookies. I compose myself, then move into the crowd. The first people I speak to are a young man and his girlfriend. What do they think about having a woman pastor, I ask. The young woman looks disgruntled. "We don't have our own priest anymore," she says. "That's sad." I ask the young man what he thinks. He says that homilies are important to him. I ask: How are hers? He says: "She's getting better."

Guliano introduces me to Sister Toni Gradisnik, the vice president of the School Sisters of Saint Francis. Gradisnik's mother is with her, too, nineteen years a member of Saint Therese's. They both love having Guliano—who was also a nun for twelve years, when she was much younger—as their pastor. Gradisnik has observed how well the supervising priest and Guliano work together, which Guliano confirms. He is open to her playing an active role in the Mass: She greets the assembly, adds to the prayers of the people, preaches, distributes Communion, and joins the procession in and out of the church. But when it comes time for the most important part of the Mass, the consecration of the bread and wine into Jesus' body and blood, Guliano must, and always does, step aside. So the ambiguity in Guliano's role does not escape them.

Gradisnik tells me about a funeral Mass they had at their convent recently when one of their sisters passed away. They wanted Guliano to serve the Mass, so she and her supervising priest arrived to do that. At that Mass, three visiting priests spontaneously "jumped up on to the altar," recalls Gradisnik, to concelebrate with Guliano and the supervising priest. But they seemed awkward around Guliano. When she did the incensing, they looked particularly perplexed. As she always does, Guliano dropped back when she had to, but this time, contrasted to the four priests who remained center stage, the diminishment of her role was particularly palpable. Instead of anger, Gradisnik felt sorrow. "My poor church," she says, shaking her head back and forth.

Marge Brey has been business manager at Saint Therese's for twenty-two years. We visit in her small office after the service. Having a woman pastor, she says "is a godsend." Brey worked with two priests prior to Guliano. One was at Saint Therese's for twenty-eight years and was beloved. But, she says, "I didn't see the compassion from him that I see with Alexandra when she's dealing with people. . . . I see it in so many things, her attitude and demeanor." When Children's Hospital used to call for the priest to visit a parishioner's hospitalized child, Brey reports, he often did not want to go. He'd tell Brey that they had their own chaplain, then refuse to come to the phone and leave Brey to make an excuse. Obviously still angry about that, Brey says, "There's no day school here. There isn't a lot for the priest to do. There's no excuse why he can't take care of something or someone who needs him." It's different with Guliano, she explains. When the hospital calls, "she'll just drop everything. She knows where she should be. . . . It did something for me to see that."

Saint Therese's parishioners have gone through a rocky transition. After losing the priest of twenty-eight years, they had another for just about eight weeks who left it to Brey to announce to the parishioners that he had resigned. Then came their first lay pastor in the person of Guliano. Acknowledging that parishioners have a hard time with a parish director, male or female, Brey thinks that had Guliano been male, she would have been more readily accepted, accorded more respect, and given more credit.

"On Sundays, after Mass, people thank the priest," Brey says. "And what does he do?" she asks rhetorically, referring to the Mass. "All he does"—and here she lifts her number two pencil—"is he says, 'This is my body, this is my blood.' " Guliano, she continues, "does *everything else*. She keeps the parish going, and there's so much." Or, Brey reports, "They say, 'Wonderful homily, Father,' even if she gives it! The priest gets all the glory." Brey thinks it's hard for some people to thank Guliano. "But it's a lot better now than it was when she first got here."

Guliano works to build community, to get people to take ownership of the parish, says Brey, which studies consistently show is the case with Catholic women pastors. "Before no one had to do anything," says Brey. "The staff did it all." Now, she reports, parishioners get the wine goblets and the host, light candles, and lock the church doors. Actually, before Guliano came, Brey used to travel from her home by car at ten each night to be sure the church doors were locked. When Guliano found that out, she told Brey, "No more. We'll get people from the neighborhood to do it." And she did. That action surprised Brey. It also surprised her how many people came forward to help. "No one had ever asked before," she says.

Guliano has a male secretary, Corby Anderson, thirty-eight. Barely a week goes by that someone doesn't call the parish assuming that Anderson is a priest and the pastor. Guliano's work impresses Anderson, and he thinks women ought to be priests. In three out of the six parishes where he has worked, he reports, three priests had to be removed for sexual misconduct—two for having sex with children under eighteen. "The Church is so adamant about maintaining this male celibate priesthood," he says. "My response is: What happens if we ordain women, God forbid, a scandal would ensue? . . . This is obviously the worst scenario we could possibly be facing right now."

Catholic Women Move into the Ministry

Guliano is a pioneer in the world of Catholic women's parish ministry. She is one of thousands of Catholic women ministers for whom the rules of engagement are still being written. They are taking on responsibilities women never had before—by invitation, by necessity, and by choice. Catholic women have also assumed major administrative posts in the Church for the first time ever—from chancellors to canon lawyers to vicar generals to chiefs of finance, operations, personnel, and administrative affairs. In the late 1990s, women held 25 percent of the Catholic Church's top diocesan posts. And John Paul II has welcomed women to certain high-level Vatican positions, in 2004 appointing the first three female members of the International Theological Commission and a woman to head the Pontifical Academy of Social Sciences, stretching the boundaries of what women can do in the Church while maintaining their rigid exclusion from the highest clerical ranks. But it is the emergence of women into every area of ministry that has presented the greatest challenge to the all-male hierarchy—particularly given its insistence on an exclusive claim to sacramental power. This has birthed a revolution on the ground, over which the hierarchy—despite its relentless restriction of ordination to men—has little control. While Catholic feminist theologians are altering the face of God, these women are remaking the face of Catholic ministry.

A major impetus for the influx of Catholic women into the ministry is the dramatic and escalating shortage of Catholic priests. In 1965, there were 58,632 priests; in 2003, that figure had dropped by 26 percent to just over

43,634. Today there are more priests in the United States over the age of ninety than under thirty. The United States has a ratio of one priest to every 1,400 Catholics, which, bad as it is, is still among the best ratios in the world, for this is a worldwide problem; in South America, the ratio is about one priest for every 7,000 Catholics, and in Africa, one for every 4,700. That reality has led some reformers to question the justice of the United States' importing priests from other countries to lead parishes; today, more than 20 percent of all Catholic seminarians are foreign-born, and the majority are preparing to be ordained for a diocese in the United States.

One result of this free fall has been the continuing increase in the number of parishes without a resident priest. In 2003, of the country's roughly nineteen thousand parishes, some three thousand were without a resident priest, with the Midwest, West, and South hardest hit (in the primarily rural Archdiocese of Dubuque, Iowa, for example, more than half of its 195 parishes—some very small—have no resident priest). Projections are that in the years ahead, that figure will rise to six thousand, or almost one-third of all U.S. parishes. This, too, is not a uniquely American problem. Worldwide, at least one-quarter of all parishes, nearly 56,000, have no resident priest.

Simultaneously, the number of laypeople involved in Catholic ministry has exploded. In the United States, nearly thirty thousand Catholics are engaged in paid parish ministry, and the vast majority of these ministers—82 percent—are female. In keeping with Vatican II's emphasis on lay participation, these Catholic women are involved in a wide variety of ministries. They head up priestless parishes—going by an array of titles, from parish administrator to parish life coordinator to the popular "pastor," though that one is greatly discouraged by the Church hierarchy. They serve as pastoral associates, directors of religious education, youth ministers, music ministers, adult education directors, liturgists, social concerns ministers, marriage preparation directors, ministers to the elderly, family ministry directors, child care ministers, and volunteer coordinators. Nuns continue to hold many of these ministerial positions, but that is changing. With the dramatic decline in their numbers (from nearly 180,000 nuns in the United States in 1965 to just over 73,000 in 2003), single, married, and even divorced laywomen—despite the Church's negative teachings on divorce—increasingly are assuming these roles.

That figure of nearly thirty thousand Catholics in paid parish ministry excludes Catholics in parish support positions—from secretary to bookkeeper, from receptionist to rectory housecleaner—who are also overwhelmingly female. Excluded as well are the legions of lay volunteers (some two hundred parishioners per parish in one study), many of whom are women. Nor does that statistic include those working in nonparish ministry—in hospitals and hospices, schools and universities, airports, seaports, and prisons—as Catholic "chaplains," another verboten job title, greatly discouraged by the Vatican. There, too, women are well-represented. Seventy percent of the 3,200 members of the National Association of Catholic Chaplains are women, as are some 45 percent of the members of the Catholic Campus Ministry Association. A study of 473 Catholic campus ministry sites found that nearly two-thirds had at least one laywoman on staff, and many had up to three.

As to the future, in the 2003–2004 academic year, almost twenty-six thousand lay Catholics were studying in Catholic seminary or college-based programs for either certificates or degrees to become what the Church hierarchy calls "lay ecclesial ministers." Two-thirds of those students were women—overwhelmingly laywomen (63 percent) as opposed to nuns (3 percent). At the same time, just 3,200 graduate-level seminarians were training to be priests. Notably, two and a half times more men were in training to be lay ecclesial ministers than priests, which may well speak to the appeal of Church service for men when it permits marriage and family life.

Navigating the Sacramental Divide

Guliano deeply appreciates the breakthrough represented by the appointment of women as pastoral administrators in the Catholic Church. But while the door to pastoral leadership is open, the door to sacramental leadership is not, which causes many Catholic women ministers a great deal of hurt, frustration, and anger.

One of their greatest challenges is navigating the sacramental divide—the Church's fault line over which the nonordained are forbidden to step. This is a struggle for Catholic women ministers like Guliano, who assails the Church for its "sexism, patriarchy, and the constraints it puts on women and married people who cannot be ordained." It requires lay ministers to call in priests—increasingly referred to as "sacramental ministers" or, more disparagingly, "circuit riders"—to preside at the sacraments, from the Eucharist to baptism, confirmation, marriage, reconciliation, and anointing of the sick—often for Catholics they barely know at all. As for the seventh sacrament, Holy Orders or ordination—of deacons, priests, and bishops—Catholic women are excluded from that ritual entirely, leading reformers to claim for Catholic women six, not seven, sacraments.

Pastor Celine Goessl is a sixty-nine-year-old nun who began to run Catholic parishes in 1976. She was one of the first women in the United States to do that work, which she continued for twenty-eight years until she retired in 2004. She's seen what issues can emerge from this model. In the twelve years that she served as administrator for Saint Luke the Evangelist Parish in Bellaire, Michigan, which ended in 2004, she reports that she "had fifty-four priests come in and out to be my sacramental minister." She attributes this to her ability to handle just about any priest the bishop sent, which did not always serve her well. Once, for Holy Week, she was sent "an Alzheimer's priest," she reports. "Oh, by Easter Sunday morning, I sat on the organ bench and could do nothing except weep because he did things so badly," she says. In addition to the Mass, adult baptisms were planned, for a man and a woman. "The man went into the pool first. [The priest] called the man by the woman's name. And I had a layperson right [there] with this priest, to point to what he was supposed to say. I had it all typed out for him. It was such a mess."

Sometimes the celebrant doesn't show up. For fourteen years, Martha Sanchez, forty-nine, married mother of three, has been the coordinator of family ministry at Holy Spirit Parish in McAllen, Texas, just stateside of the Mexican border. Sanchez manages the child care program, bilingual religious education, the annual multicultural festival, and the celebration of quinceañeras—the Mexican coming-of-age rituals for fifteen-year-old girls.

Sometimes, the priest or deacon scheduled to preside at the *quinceañera* service forgets to come, reports Sanchez. Knowing how devastating it would be to the glowing girls and their families to cancel the event, a layperson will step up and lead. "I have done it, and other people have done it," Sanchez says. "This is the kind of parish we have. We always do everything for the community."

The failure of a priest to show up happened frequently enough at Communitas in Washington, D.C.—one of the country's oldest Intentional Eucharistic Communities, a type of small faith community—that it prompted a period of soul-searching among its members. Communitas' model for their Sunday Eucharistic service is to have both a priest presider and a lay presider. One Sunday, they waited and waited for the priest. When they felt they could wait no longer, Sister Alice Zachman—a devout Catholic who has felt a call to the priesthood since she was a child, who knows the Mass and has concelebrated with priests at Communitas in the past—went ahead, with the group's support, to begin the service. Suddenly, the tardy priest arrived. He swept into the room, quickly vested, waved an astonished and wounded Zachman aside, and said the Mass himself. It was a devastating moment for the community, witnessing the humiliation of a woman of Zachman's spiritual stature for no reason but her gender. As a result, the members decided that, in the event the scheduled priest is unavailable to co-preside (either cancels at the last minute or fails to arrive on time), they will go ahead with their lay presider—man or woman—and celebrate the Eucharist themselves.

Some communities of Catholic religious sisters claim the right to celebrate Eucharist privately, for their members. Janet Walton is a member of the Sisters of the Holy Names, a cofounder and participant for more than twenty years in the women's New York Liturgy Group. She is also a professor of worship at Union Theological Seminary in New York. In 2002, she and hundreds of members of her order at their annual meeting together celebrated the Eucharist. Walton admits that "I could be called a heretic because of what I do Eucharistically." As she explains her thinking, if someone were to say to her that she could not celebrate the Eucharist because she is not a member of their community, not a good leader, or has no gift for ministry, she could accept that. "But if the only reason is that I am biologically different . . . from a man, physically different, then I can't accept that," she says. "I know a lot about what the Eucharist is supposed to achieve. I study it. . . . I teach it. I lead it. . . . I believe that gathering around food is one of the very basic commitments that Christians have. . . . I think that women have to stand up and say, 'I have the authority to do this.'"

The Mass v. the Communion Service

Lay ministers, including women pastors, are officially forbidden to celebrate the Eucharist or Mass, but in some places they are permitted to lead so-called "Communion services." These services, in turn, have become essential in many parishes that have no resident priest and would otherwise have no Sunday service at all. Officially called "Sunday Service in the Absence of a Priest," Communion services feature Bible readings, a homily (or sermon), and the distribution of Communion hosts consecrated by a priest at a previous Mass.

Pastor Celine Goessl conducted many Communion services during her twenty-eight years of pastoral ministry. In fact, not only did she do them, but she was also incredibly proud to have trained five laypeople and another nun to do them as well. That act was in keeping with her pioneering commitment to the preparation of laypeople to take over priestless parishes, which was the subject of her dissertation in the 1980s. When she pastored Saint Luke's in Michigan, they used to have a Communion service every Wednesday night, rotating leadership among the seven people, who were "in charge of everything." To Goessl, it was "wonderful."

Carolyn Groves, a theologian and director of religious education at Saint Athanasius Church in Evanston, Illinois, applauds Communion services for the way they open the sacramental celebration to the people. "With Eucharistic services, the focus is on the priest," she observes. "But with Communion services, the focus is on the community."

Another fan of Communion services is Celeste Anderson Byrne, the forty-two-year-old president of the board of Mary's Pence. It is a nonprofit organization (whose name is a take off on "Peter's Pence," the pope's annual fund-raising appeal) that works throughout the Americas to support women's empowerment through Catholic-women-led grassroots ministries. Byrne is an African-American Catholic; she and her husband attend a small Catholic parish, Saint Williams in Louisville, Kentucky, that has a woman pastor, as well as a local Baptist church. At the Catholic Church, Byrne enjoys what she calls "the bread service." "I know some people are like, 'Well, it's not a real Mass.' Yes, it is. We're all gathered. We're praying. We're eating. . . . For me, it's the same."

But the very fact that a Communion service may appear to be a Mass is deeply distressing to the Catholic hierarchy. So disturbing was it to the Kansas bishops that in 1995, they issued a searing statement restricting Communion services to emergency situations. "The priest is not just a functionary who consecrates the Eucharist, pours water, anoints with oil, or absolves the penitent . . . not just a circuit rider who offers Mass and celebrates the sacraments," they wrote. Among the trend's "disturbing implications," they listed a "blurring" of the lines between the ordained and the nonordained and between pastoral ministry and sacramental ministry, as well as a "blurring of the need for priests," which is obviously to them the most frightening thing of all.

Pope John Paul II has expressed similar alarm. In his 2003 Holy Thursday Encyclical on the Eucharist, he shared his deep and moving reverence for the Eucharist. While finally, and begrudgingly, acknowledging the devastating "scarcity of priests," he failed to credit the efforts of lay ministers leading Communion services and instead seemed to scold them. "In some places," he wrote, "the practice of Eucharistic adoration has been almost completely abandoned," as if this has happened by some neglect of the laity and not because there are not enough ordained men to go around.

The pope expressed his "profound grief" at new ways and places of Eucharistic celebration, in the absence of an ordained minister, celebrated "as if it were simply a fraternal banquet." In what can only be read as an insult to devoted Catholics who participate in these services, the pope concluded, "The Eucharist is too great a gift to tolerate ambiguity and depreciation. It is my hope that the present Encyclical Letter will effectively help to banish the dark clouds of unacceptable doctrine and practice."

On the other side of the divide, some progressive Church reformers worry that with the proliferation of Communion services, Catholics are being deprived of what is so central to the faith, participation in the Eucharistic consecration. Sister Christine Schenk of Future Church says, "The foundational theology of the Eucharist is of a community gathered being the ones to consecrate." When the gathered community does not have an opportunity to share in the consecration, she continues, "That's a corruption and a regression of the Catholic understanding of what that meal is about, which from its inception has been done in memory of Jesus."

As a result of its concerns, the Church hierarchy is becoming increasingly desperate in its proclamations aimed at maintaining the sacramental divide. In the early 2000s, the Vatican, and then the U.S. Conference of Catholic Bishops with the Vatican's approval, issued new instructions concerning the Eucharist and Mass, which clipped the wings of lay ministers. Those new instructions forbid Eucharistic ministers—Catholics who volunteer and are trained to help with Communion—to walk with the priest in the entrance or exit processions; to be on the altar, except after the priest receives Communion; to touch the chalice, until the priest literally hands it to them; to break the bread; or to pour wine from one chalice to another. It was only with special permission from Rome that Eucharistic ministers in the United States were allowed to drink leftover wine or to even "purify"—that is, to *wash*—the sacred vessels. In 2003, the Vatican was working on a draft document on liturgical abuses that in its earliest incarnation would have banned female altar servers once again, reinstated Communion rails separating the clergy from the people, and forbidden clapping or liturgical dance.

"The saddest part of all," says Schenk, "is that the focus is so messed up. It's a focus on minutiae. Instead of building up the people of God and talking about who we're called to be in this world where we have war and terrorism and whole countries dying of AIDS and starvation, they're focusing on who can wash the Eucharistic vessel." She finds it hard to imagine "that people are sitting [in Rome] worrying about these kinds of things." To her, "it's a sign that they're recognizing that they're losing power." The increasing rigidity has filtered down to the local level as well; in 2004, diocesan authorities in New Jersey declared invalid an eight-year-old girl's First Communion because a sympathetic parish priest had dealt with her rare digestive disorder, which precludes her eating wheat, by giving her the Eucharist in a rice wafer form.

More Minefields for Women Ministers

The problem is not only that laywomen ministers are forbidden to celebrate the Eucharist. They also must navigate other sacramental divides. "Regardless of how much time I spend preparing couples for marriage or families for the funeral rites, my role at the liturgies of marriage and funeral Masses is limited by canon law," Guliano wrote. "I have spent hours at the bedside of a suffering person, held their hands, counseled their family, and asked the hard questions around the death and dying process. But when the persons request the sacrament of the anointing of the sick or the sacrament of reconciliation, I cannot administer the sacraments to them."

Ellen Radday, former regional director of the National Association of Catholic Chaplains and a member of Communitas, tells the story of being the Catholic chaplain on duty late one night when a man found unconscious in his home was rushed to a Maryland emergency room. The man, near to death, needed anointing, but only a priest can perform that sacrament, and the on-call priest was sick. "I stood by [the man's] bloody body with my hand on his shoulder," reports Radday, "and I prayed with all my heart, good solid Catholic prayers—the Creed, the litany of the saints, an Act of Contrition, the Lord's Prayer—whatever might give him comfort and a spiritual connection if he could possibly hear me." After that, Radday left a voice mail for the patient's parish pastor. He came by the next day but let the chaplaincy director know how annoyed he was by the late-night call.

Hospice chaplain Robin Imbrigiotta describes what it is like to be forbidden to anoint. At forty-nine, she is the only full-time chaplain and the first Catholic with the Hospice of the Visiting Nurse Service of Akron, Ohio. "There've been situations where someone has died, and the priest didn't get there in time," she says. "The family is standing around the bed, and I'm comforting these people, and we're talking about the patient. The priest walks in, takes off his hat, looks at everyone, and says, 'Oh, I'm so sorry.' And they'll say, 'Father, say a prayer.' And Father will say a prayer . . . then walks out. They're like, 'Oh, thank God Father was here.' "

"Personally, from a very egocentric place, I feel as though I'm not meeting their needs. But I know in a greater way, I am because I'm offering comfort and compassion to them. But Father offered them the Church. I guess I wish I could offer them the Church, the Church with its heart, the Church with its hope. I can offer me . . . but I cannot officially offer the Church and that hurts. It hurts."

For these reasons, the National Association of Catholic Chaplains has urged the U.S. Conference of Catholic Bishops to allow Catholic lay chaplains to anoint dying patients. Although no such permission has been granted, there have been increased opportunities for Catholics to be anointed after Mass rather than at the precipice of death, along with a concurrent move—welcomed by people like Guliano—to remind Catholics that the real last sacrament is Communion or *Viaticum* (food for the journey), which a lay chaplain can provide with a previously consecrated host. While these efforts may distract attention from the lay chaplain's inability to anoint a dying Catholic, they certainly do not solve the sacramental problem.

Chaplains are also technically forbidden to provide the sacrament of reconciliation. While a master's of divinity student at Union Theological Seminary in 2000, Katie Bergin, thirty-one, did a pastoral internship as a chaplain on the oncology floor at Georgetown University Hospital. When it came to anointing the sick, patients often wanted to see a priest. That worked out fine for Bergin; the Jesuit she worked with appreciated her relationships with her patients and "wanted me in the room to say the prayers with him. . . . We did it together," she reports.

Her experience with the sacrament of reconciliation was different. "I heard a lot of confessions, but I never heard anybody say, 'Will you send the priest so I can really go to confession, for real?' " she says. "I had people ask me if the conversation we just had was confession. . . . I would always ask them, 'What do you understand confession to be, and what are you looking for?' " It is Bergin's belief that as "another human being, I can be a vehicle to help you experience God's forgiveness. I don't have to have a penis to do that."

And then there is baptism. Sister Goessl recalls a Sunday morning when the priest who was to preside for her called in sick. She reminded him that not only a Mass but a baptism was scheduled for that morning. Deeming conduct of the baptism "a case of necessity," the priest told Goessl to do it herself. She did. "Of course, priests from the area found that out, and they questioned whether this was a valid baptism," reports Goessl. "I said, 'What's the matter with these guys? Send them back to school to learn a little theology. Anyone can baptize in the case of necessity.'"

Weddings, where "the priest's role lasts five minutes," says an exasperated Guliano, is another area of disputed terrain. The couple marries each other, but the priest must pronounce them wed, so laywomen ministers are prohibited from officiating at Catholic weddings. Despite this, I've talked to nuns who did marry people when asked, but on the q.t. As far as funerals, if there is no Mass, Catholic women ministers are able, within church law, to lead those services—and they do. And some Catholic laywomen trained in theology, as well as some priests who were forced to leave the priesthood to marry, have joined the Federation of Christian Ministries, through which they can become certified Christian ministers legally and spiritually empowered to conduct weddings and other Catholic-style services. A Catholic advertising executive, Louise Haggett, provided another alternative to Catholics suffering from the priest shortage. Unable to find a priest to minister to her dying mother, Haggett, in 1992, founded CITI (Celibacy Is the Issue), or its jazzier title, "Rent-A-Priest," a national network of more than two thousand married Catholic priests available to conduct weddings, funerals, and other religious services for Catholics.

Catholic Laywomen in the Ministry

Catholic laywomen in the ministry, as opposed to religious sisters, have their own special challenges to deal with. Jacqueline Landry, forty-six, is Harvard University's first Catholic laywoman chaplain, a job she took in 1994. She differentiates herself not only from priests but from nuns, too. "All my energy is spent . . . negotiating . . . door number three . . . Father, Sister, me," she explains. "I'm about this third way. If you're a nun, you have to give something up to be perceived as having a little bit of power. . . . You have to give up sexuality. You have to be perceived as a neutered, neutral woman." Now, not all the nuns, she adds: "I've worked with Benedictines who wore bigger earrings . . . more lipstick than I probably would in my lifetime."

From the start of her career in the Church, says Landry, she found herself "a lightning rod for issues around sexuality." It is her experience that many women ministers deal with that, with "what to wear in the pulpit . . . lipstick or not? . . . High heels or not? If we're viewed as sexualized, what does that mean? . . . How do you negotiate the props of being women in our culture and at the same time [be seen as] spiritually enriched people?"

The fundamental problem, as Landry explains it, is the difficulty the Church has integrating sexuality with spirituality. "Why can't we say that God comes to us through our sexuality?" she asks. Why must sexuality be seen as "a little problem we've got to negotiate around?"

Because women have represented sexuality in the Church, laywomen ministers like Landry bear a special burden. Back in her days as a campus minister in the Midwest in the early 1990s, she was invited to preach one Sunday at the University of Minnesota's campus church. When the liturgy ended and she was shaking hands with the parishioners, a visitor to the campus parish came up to her and spat in her face. "You are what's wrong with this Church," the man fumed, "you feminists who want to be priests." "I don't want to be a priest," responded Landry. To which the man, perhaps getting to the heart of the matter, snapped: "Why aren't you covered up more?" In fact, it was the dead of winter, and Landry was covered, head to toe. "I realized, 'Oh my God. He just had his paradigm rocked. He's coming to church, where he has put sexuality safely outside those doors, [and] he encounters me, a woman." She also realized that the only part of her that was exposed was her mouth.

Preaching: The Eighth Sacrament

Preaching is an extremely controversial area of ministerial practice for Catholic women today. Canon law and Church regulations forbid a layperson from preaching the homily, which is the sermon that follows the Gospel reading at Mass; that role is restricted to an ordained person. However, canon law does permit lay Catholics to preach at other times, and they do;

many pastoral administrators have authorization to preach at Communion services, while some lay Catholics even preach at Sunday Mass, as long as a priest or bishop gives special permission or makes an introduction, naming the layperson's sermon not a homily but a reflection.

Jacqueline Landry has preached at Saint Paul's Parish—the lower church, home of the Catholic students' association—on Harvard's campus. She describes the university environment as increasingly conservative, a challenge for a lay Catholic woman in a nontraditional role. And a lay Catholic woman is all Landry wants to be. While some Catholic female chaplains vest for their work, Landry does not. "I don't want to look like I'm trying to be a priest," she says. "I want to just look like a woman who wears DKNY panty hose . . . a woman out of the pew . . . [whose] power is her baptism, not her clergy power." She wants to project "an ordinary holiness, a holiness that you can have married—I'm not married, but you could be married. You could have just a regular job. And your spirituality is as deep and as important as a nun or a monk in a monastery."

As for preaching, Landry reports, "It's tricky how it's done." The priest has to "kind of frame what I'm going to say." But, she adds, "to the people in the pews, it looks like a homily." For her part, says Landry, "I don't care. The point is, when I'm up there, there are six hundred students at this Mass I'm thinking of, and I am the first woman they have ever seen open her mouth. And you should see their looks. It's as if they're caught in headlights."

Women preachers not only change the image; they change the subject. Svea Fraser, the first female Catholic chaplain at Wellesley College as well as a founder and board member of Voice of the Faithful, recalls her first sermon, in the mid-1980s. It was a requirement for a homiletics course she was taking at Andover-Newton Theological School. She planned to give it at a Protestant Church until the priest at her Catholic Church, Saint John the Evangelist—later the birthplace of VOTF—found out. With great enthusiasm, he invited her to give the sermon at Saint John's.

The date was to be the second Sunday in Advent, and the Gospel reading was the Proclamation of John the Baptist (Luke 3:1-6). Fraser went to a little chapel to ponder the passage about "the voice of one crying out in the wilderness: Prepare the way of the Lord." While praying, she noticed a banner bearing an image of Mary as "a woman great with child." "I just burst into tears," says Fraser, who knows a lot about motherhood as the eldest of ten children and mother of two daughters. "My God, I thought, Advent is about giving birth, bringing something to life. You don't know quite what it is, but it's inevitable that it's going to happen. . . . That's what 'preparing the way' is all about." But she also thought, "I can't preach about that! . . . First of all, there would be a woman preaching. . . . Then, to talk about something that was so feminine in a Catholic Church, birth, being a mom. . . . I'd really be hitting them with feminine stuff. And I'm a coward anyway," she adds. "I march *behind* people."

Fraser shared her trepidation with the black Baptist minister who was her homiletics teacher. "I can remember his looking at me and saying, 'What do you really want to do, Svea?' I said, 'I want to preach.' Then he came over and put this big, black hand on my shoulder and said, 'Well, *preach it, sister.*'" Fraser did, wearing a pink suit, to a full house and a round of energetic applause. Later, she reports, the priest who let her give the homily was reprimanded for doing so by Cardinal Law.

A layperson gets to preach at least once a month at Celeste Anderson Byrne's Louisville, Kentucky, parish, and it's usually a woman. With a master's in pastoral studies from the Aquinas Institute of Theology, Byrne has gotten to preach six times so far and loves it. She reports that they don't pay attention to the rule book. "We call it preaching," she says. "I've been to other parishes where . . . the Father will get up, and he'll do his little introduction and say, 'Okay, technically this isn't a homily, but we're going to hear from da-da-da today.' We don't do that. No. I get up. I read the Gospel. I preach."

Communitas' services are held in a comfortable, living-room-like space belonging to the Catholic homosexual rights group Dignity, in Washington, D.C. There is no pulpit, no altar, no perch from which the ordained gazes down to the quiet and obedient nonordained. After the Gospel reading, different voices rise spontaneously from around the room in what is called a "dialogue homily."

While Catholic women preaching can require rule-bending and a certain amount of risk for the priests who invite them, deacons in the Catholic Church officially preach. The number of ordained deacons—who, like priests, must be men—has increased in the United States from under 1,000 in 1975 to 13,600 in 2003. Deacons can also baptize, witness marriages, and conduct funeral services. A deacon can be married, but in a rather gruesome nod to the Church's legacy of misogyny, he has to vow not to remarry if his wife dies. Deacons are glorified by the Church as sacramental ministers, even though they don't require the full-time commitment to their work that is required of the legions of Catholic women in the ministry. At Voice of the Faithful's Northeast regional conference in 2003, a deacon with a New York area parish gave me his Benjamin Moore Paints business card. It listed him as a sales representative but also had his deacon address scribbled on it. Without special permission or fanfare, he can get up anytime, preach the Gospel, and give a homily. None of the women you're reading about here can do that.

That state of affairs led Phyllis Zagano, author of *Holy Saturday* and an otherwise ultraconservative, self-described "right-wing" Catholic, to argue passionately for the restoration of women deacons in the Catholic Church. She reports that the Church had ordained women deacons until the fifth century in the West and the eleventh century in the East, and there is no reason we shouldn't have them now. Illustrating her position, Zagano told a *Pittsburgh Post-Gazette* reporter, "Why not take the woman in the soup kitchen who has lived the Gospel and let her get up at noon Mass and talk about it? We need people in the ministry of the word who know more about the Gospel than about golf."

Taking advantage of the leeway that does exist, Patricia Hughes Baumer, a Catholic who holds a master's in divinity, and her husband, a former priest, cofounded the Minnesota-based nonprofit Partners in Preaching to train lay preachers at the parish and diocesan levels. In the Saginaw diocese in Michigan, the late and beloved Bishop Kenneth Edward Untener authorized associate pastors and pastoral administrators in the diocese to have homiletics training and deliver reflections at Mass. And individual Catholic priests invite laypeople to preach but do so quietly, to avoid the wrath of disapproving superiors and colleagues.

Still, there is a lot of resistance from the hierarchy to lay preaching. In his 2004 Easter Week Mass for Boston area priests, Archbishop Sean P.

O'Malley focused on preaching. Noting that preaching has been called "the eighth sacrament," he pronounced it the "first task" of the ordained priest, and without actually excluding or denigrating the attempts by lay Catholics to preach, he declared the priest's preacher role to be "irreplaceable." In spring 2004, the Vatican issued a document declaring in no uncertain terms that lay preaching should only be allowed "when necessity demands it," should not be seen as "an advancement for the laity," and must never be "transformed from an exceptional measure into an ordinary practice." Ironically, some bishops may have learned to preach in a homiletics course taught by a Catholic woman at a seminary, yet remain silent as the Vatican tightens the restrictions on Catholic women's right to preach.

Young Women Ministers for Social Justice

Many Catholic women—especially young Catholic women—are drawn to social justice work. They take their longing to minister with them to their jobs in nonprofit organizations, where they combine their deep Catholic faith with a burning commitment to peace and justice. They remain devoted to the Church's social principles, even though the institution refuses to ordain them or even to recognize their independent ministries. Sara Willi and Kelly Jones are two recent college graduates. They live in a house in Washington, D.C., that is owned by a couple who rent out rooms at reduced rates to people doing social justice work. They traveled together to Call to Action's 2003 national conference. Both are deeply devoted to the Catholic Church and see their jobs as their ministries.

Willi is a stunning twenty-four-year-old woman with black hair, blue-gray eyes, and an earring in her right eyebrow. She says she began her social justice ministry at the age of fourteen, when she became a vegetarian. Later, she volunteered at soup kitchens and homeless shelters and studied in South Africa and Northern Ireland. At the College of Saint Benedict in Minnesota, she majored in peace studies. There, she says, "I claimed my faith for myself." She also began to see the Church's discrimination against women, in male-dominated prayer language and in worship services where the college's independent nuns "could just do so much," then had to bring in a priest. Deeply committed to getting women's voices heard, Willi tried, unsuccessfully, to bring *The Vagina Monologues*—Eve Ensler's brilliant play about female sexuality and the cost to women of its demonization—to campus, insisting that "God gave us vaginas, and we should be able to talk about them!" Nuns were among those who supported her effort.

Willi works for the Education for Peace in Iraq Center. She joined the center right after President George W. Bush's speech declaring Iraq, Iran, and North Korea the "axis of evil." She describes the center as working to educate people about Iraq, to bring the faces of Iraqis to America, and to end the humanitarian crisis there. Today, she says, "My calling is to do peace and justice work." Her faith "keeps me grounded and hopeful."

Willi's roommate is twenty-six-year-old Kelli Jones. She is a development assistant at the Center of Concern, a Jesuit-founded independent organization committed to global economic and social justice, where former nun Diann Neu organized a major conference on women in the Church more than twenty years ago. Jones went to Georgetown University School of Foreign Service, majoring in both peace and women's studies. Then she joined the Jesuit Volunteer Corps. Jones spent two years teaching in an alternative Catholic school in Belize. While there, she was confronted with tremendous poverty and the struggling lives of Belize women, who she describes as "neither recognized nor honored for the role they play in holding the whole place together." Jones taught Christian social justice, telling her students "everything about respecting women . . . human dignity . . . human rights." She taught them about the courage of martyred San Salvadoran Archbishop Oscar Romero, "about sweatshops. I talked to them about communities, about celebrating faith."

She sounds like a minister. Sitting with her in a dim corner of the nearly deserted lobby of the Milwaukee Convention Center, I notice that Jones is wearing a purple stole, the symbol for women's ordination. I ask if she has a call to be a priest. She tells me she attended a prayer service that morning, where a woman asked the same question. Jones's reply to the woman was "maybe," to which the woman responded by giving her a purple stole. Then, reports Jones, "a group of women gathered around me and laid their hands on me and told me that they were commissioning me to go out into the world." Since then, "I've spent a lot of today crying, but in a joyful way. I feel

I'm just full of grace and light . . . illuminated." It's the only validation a young Catholic woman like Jones can get right now for her commitment to Catholicism, social justice, and independent ministry.

The Field Develops but Tensions Mount

The U.S. Conference of Catholic Bishops is addressing—through panels, papers, surveys, and conferences—the issues raised by this new field of lay ecclesial ministry. According to insiders like Sister Amy Hoey, project coordinator for the U.S. Bishops' Subcommittee on Lay Ministry, progress is being made. They are working to define lay ecclesial ministry better. They are examining ways to improve the relationship of ordained ministers to lay ministers, with an emphasis on collaboration. They are supporting the development within dioceses and various professional associations of appropriate certification, education, and training requirements for lay ministers, with the recommendation that dioceses—particularly in poor communities—make more funds available for that training.

The bishops are suggesting recognition for lay ecclesial ministers through rituals of installation or commissioning, such as the commissioning ceremony that was held by the archdiocese of Los Angeles in 2003 for eighteen pastoral associates (fourteen of them women) who had completed its new training program. And they are addressing the personnel and human resources issues that have long plagued lay ecclesial ministers—from inadequate wages and benefits to lack of contracts, to no job security, especially when a new pastor comes in. That particular matter came to a head at Holy Spirit Parish in McAllen, Texas, in 2003. On the new pastor's first day on the job, Martha Sanchez and the three other women ministers—including the indefatigable Ann Cass, who has been running things at Holy Spirit for nearly twenty-five years—arrived at work, as usual. Suddenly, with no notice, they were all summarily fired. They were ordered on the spot to leave their jobs, as security guards patrolled the halls and a locksmith changed the locks on all the doors.

Holy Spirit was the first parish in the United States to be unionized, which it did under the auspices of the United Farmworkers of America. The fired women rallied the media, their parishioners, and the community. Hundreds met and marched. Some parishioners refused to enter the church, holding services outdoors, on church grounds. Despite Brownsville Bishop Raymoundo Peña's union-busting tactics (he reportedly threatened to withhold funding from other parishes that were considering unionizing), the women and the union prevailed. Peña was forced to reinstate them in their jobs.

"What happened to them isn't unusual," says FutureChurch's Chris Schenk. "That happens, I'm afraid, all of the time. I have a whole folder of women around the country who've been similarly dismissed. What is unusual is that [in Texas], they decided to take a stand and resist it." She adds that "most people do not want to make a fuss because they're very afraid they'll be blackballed in that diocese and will never be able to get a job in any parish anyplace else. . . . Usually the presumption is that the ecclesial minister did something wrong or they wouldn't have been let go. It's a very unhealthy system."

Theology professor Luke Timothy Johnson shares that concern about the treatment of Catholic women in the ministry today. "Everyone knows that most Catholic parishes in this country would close up tomorrow if it weren't for women," he wrote in *Commonweal* in 2003. "[They] are carrying out most of the work of ministry in many, if not most, parishes. The same abuse of power with which the male clergy exploited but never fully honored the ministerial labors of vowed religious women . . . is now being perpetuated in the exploitation of single and married women in local parishes. . . . An increasing number of American Catholic women do see the pattern, and they are angry. . . . [I]f Catholic women finally get angry enough to walk out, the game is close to over."

That lay ministry has a female face has not escaped the notice of the bishops. "In our conversations and consultations, not just with the bishops but with other leaders in this field, there's a concern about the feminization of the ministry," says Sister Amy Hoey. I'm stunned to hear her say that, given the all-male composition of the body for which she works. "I think they'd like to see a more balanced representation," she explains, seeming blind to the irony. When I ask about that, she says: "One imbalance is not corrected by another imbalance."

Clearly, there has been backlash. Some dioceses have begun to replace women pastors with male deacons. Retired pastor Celine Goessl, who was active in an association of Midwest pastoral administrators, reports that the pressure for that action came directly from Rome. In 2004, in Lexington, Kentucky, Bishop Ronald Gainer fired five people, including four of the diocese's senior women leaders—the directors of pastoral services, parish leadership, lay ministry formation, and educational ministries, as well as the male chief financial officer. The women were regarded as "strong advocates for greater lay involvement in church leadership" and founded a program called the New Faces of Ministry. Reportedly, pressure from conservatives in the diocese contributed to the firings. The diocese said that it planned to reorganize; at the time of the firings, it had established two new offices, both headed by ordained male deacons. In some dioceses, women chaplains report attempts to split their role, taking all spiritual and sacramental responsibility out of their hands and putting it into the hands of priests, while delegating to the women the administrative work alone.

In an effort to ensure that the lay ministers are conservative on Church teachings, Bend, Oregon's Bishop Robert F. Vasa ordered lay ministers to sign an "affirmation of faith," condemning a host of sexuality issues, from masturbation to premarital relations to abortion to homosexuality, or lose their jobs. Lay liturgical minister and cantor Wilma Hens refused to sign and publicly quit her job, along with at least six other lay ministers.

Some see the hierarchy's animosity toward women's growing power in Catholic ministry as so strong that the hierarchy would prefer to see a shrinking Church rather than bring more women in. As a case in point, in 2004, Boston Archbishop Sean O'Malley announced the closure of more than one-fifth of the area's 357 parishes, which he attributed to financial woes, caused in part by the clergy sex abuse crisis, as well as to the critical local shortage of priests and declining Mass attendance. Boston College theology professor Pheme Perkins saw another, more fundamental influence at work. At the 2004 conference on "Envisioning the Church Women Want" at Boston College, she declared publicly: "I think the current parish closures are an attempt to throw women out of where they could go, which would be to lead parishes without priests."

In 2003, for the first time in nearly twenty years, there was a dramatic drop in the number of lay Catholics entering lay ministry training programs, to just under 26,000 from more than 35,000 in 2002. Two-thirds of the students, however, remain women. One factor contributing to the decline may be a diminishing number of such programs. While the trend among religious order schools (e.g., operated by the Jesuits) has been "wholeheartedly toward incorporating lay students," the trend among diocesan seminaries has been "downplaying" the presence of lay students or even eliminating the lay students altogether, according to Sister Katarina Schuth, a Catholic seminary professor and expert on Catholic theological education. Other reasons that have been put forth for this sudden and precipitous decline are the sex abuse-crisis, which may be turning Catholics away; the workplace treatment of Catholic women ministers; a diminishing number of jobs; and the difficulty women ministers encounter in working with new priests, many of whom are conservative and not open to women's participation. According to Schuth, an estimated one-third of Catholic seminaries operate closed systems, training seminarians in complete isolation from lay students—including women. That can only reinforce the reluctance women ministers report among some new priests to work with them at all.

The final results of the pushes and pulls in the area of Catholic women's ministry in the Church remain to be seen. For now, thousands of Catholic women are active in the ministry, "getting into people's imaginations," as chaplain Katie Bergin put it, providing Catholics with a new vision of the sacred. In the process, they are altering the very concept of sacramentality and its relationship to the people in the pews.

Loyola University theology professor Susan Ross interviewed many Catholic women in ministry for her book on a feminist sacramental theology. She found that "women are already acting as sacramental ministers in the Church, although it's not official," she told me. She sees the Church today as "kicking and screaming like crazy to keep women as far away from the sacred as they can . . . trying to draw these lines [in a] desperate action to keep women out of the sanctuary, so to speak."

In practical terms, women are moving sacramentality from its association with the priest's "magic fingers" and "magic words," as one Catholic woman chaplain put it, to an emphasis on participation, connection, and relationship. That, in turn, is putting sacraments in the hands of the community. "What I see in the broadening of this definition of sacrament is an intent to see sacraments as actions *of the community*," wrote Ross. To her mind, "sacraments are not purely priestly actions; they are not restricted to the actual moment when the sacrament is 'conferred.' They are linked to the ongoing process of recognizing God's presence in all of life and most particularly, within the community."

Corpus Christi Parish in Rochester, New York, was one such community. Out of it came a woman who dared—in the company of thousands—to cross the final sacramental divide. It earned her the title "the Rosa Parks of the Catholic Women's Movement."

13

✠

ONE WOMAN WHO REFUSED TO WAIT: THE ORDINATION OF MARY RAMERMAN

I plant myself in the lobby of the Eastman Theatre in Rochester, New York. It is the morning of November 17, 2001, and freezing cold outside, but the sun is shining, and the sky is clear. People arrive in droves. Finally, the interior double doors to the three-thousand-seat auditorium swing open. The furtive crowd pours in. "It's a good day!" one shouts. "No, it's a *great* day!" shouts another.

All around people shake hands, hug, kiss, laugh, and really laugh. Kids skip up and down the aisles. Among the diverse crowd are priests incognito, Catholics defying the local bishop, interfaith ministers, fathers and sons, mothers and daughters, and women in white robes, some comparing what is about to happen here to Afghanistan's challenge to the Taliban. There are Catholics trying to hide from journalists like me, and others racing to get a seat. The chorus members make their way to the bleachers on the giant wooden stage. The saxophonist, the harpist, and the drummer warm up, filling the air with skittish notes. The place is so alive that you almost have to cut your way through the energy in the room to get from one end to the other.

Standing in front of his seat, Victor Schwarz observes the clamor. "I followed the saga of Spiritus Christi from afar, ever since the whole crisis a couple years ago," says Schwarz, a retired lawyer who drove all the way from Reston, Virginia, to be here. "I know lots of parishes where the same thing happened. Basically, you had a really vibrant community led by a liberal priest. The bishop would come in; the priest is gone; and everybody let it happen. For the first time, at Spiritus Christi, the people stood up and said: 'No.' . . . My experience of the Catholic Church has been that nothing is going to change unless people are prepared to defy authority." Schwarz also thinks this could be "an historic event." He characterizes it as a "high-profile public statement," not just "some fringe thing" being held "off to the side." His hope is that, by seeing the people of Spiritus Christi do this, other people will find the courage to see that "we can do it, too."

The woman of the moment is Mary Anne Whitfield Ramerman. It strikes me as significant that her name is Mary. Not the passive, anything-goes, I'm-just-a-vessel Mary—but Elizabeth Johnson's Mary. The emboldened Mary of the Magnificat, the Mary who found her voice and said "yes" to God. I have waited my whole life for this day or something like it, the day when I would get to see a Catholic woman ordained.

Out of Paradise, into the Future

Mary Ramerman was not always a rebel. In fact, she wasn't always a Catholic. Born in 1955, she grew up Methodist in California. She fell in love with Catholicism deep in the woods. She got involved with the Roman Catholic Church at the age of sixteen, she tells me in a soothing voice that has a bit of a trill. She was looking for a way to get out of the house that summer. So she signed up to work at what turned out to be a Catholic summer wilderness camp. Taking advantage of the liberalization of Vatican II, Masses were held among the redwoods. "That was my first introduction to Catholicism, and it was really a beautiful introduction," Ramerman says.

She went to Humboldt State in California, with a lot of counselors and others who had led that Catholic camp. That's where she met her husband, Jim Ramerman, a Catholic. At the age of twenty-five, shortly after they married, she converted to Catholicism. They both became youth ministers in several parishes, then worked their way up to diocesan posts. It was a natural progression for Ramerman, who wanted to be a minister from the age of seven. She also went on to get her master's in theology. But a few years later, the two became disillusioned with the diocesan bureaucracy, especially its treatment of its employees, and began to look for other jobs. One offer came from a small struggling parish, Corpus Christi, located clear across the country, in the slums of Rochester, New York.

"We came out here for a week to check it out," recalls Ramerman. She and her husband were being considered as a team to provide family ministry and to coordinate other ministries. They left balmy California in early May and found snow on the ground in Rochester. They found a lot of people hanging out and drinking beer on the curb. They also found a ramshackle house. A donation to the parish, it was covered in an ugly green, from the exterior to the inside walls to the furniture to the kitchen cabinets. And everything that wasn't green was dilapidated. "You have got to be kidding," said Ramerman to her husband on Monday night. "I'm leaving tomorrow morning."

She didn't. As soon as the Ramermans met Corpus Christi's little staff of five and saw their work, their doubts faded. First, they really liked the pastoral administrator, Father James Callan. An exuberant, outgoing priest, Callan had given the hierarchy pause from his earliest days in the priesthood. In his book about his seminary days and the Corpus Christi ordeal, *The Studentbaker Corporation: A Vehicle for Renewal in the Catholic Church* (a send-up of a student-financed Studebaker), Callan reports he was nearly expelled twice. He had refused to wear a Roman collar and then declined to live in the cushy priests' rectory, served by the female help. A Vatican II priest, Callan believed in living a humble life and serving the poor. He also believed in an inclusive church, open to gays and lesbians; in welcoming all people to the Eucharistic table; and especially in the development of women's ministry and women's ordination.

Callan, who assumed the reins at Corpus Christi when he was only twenty-nine years old, had been working tirelessly at launching parish ministries and promoting parish involvement, which won over the Ramermans. Recalls Mary of their week in Rochester: "One day we went into Attica prison. Another day we went into the health center, [where] a man came in whose wife had just died of sickle-cell anemia. Another day I met with forty people from the parish [who] wanted to know who we were. At the Thursday night folk Mass, on one side of me was a homeless man; on the other side was a woman running for city court judge."

But the parish was dirt poor. Staff lived out of an envelope of donations. Even asking for a salary seemed excessive, under the circumstances, but an apologetic Ramerman did. The couple settled on $6,000 a year, made a commitment for two years, and in the summer of 1983, packed their two-year-old son, Matthew, into their red VW Rabbit and drove out of paradise into the future.

Building a Ministry

Mary Ramerman's hair is a lustrous auburn. Her eyes are true green. She locks you in with her gaze, so you don't care, really, if you ever get out. She exudes warmth and generosity. She also has a big, bold, body laugh, as well as nice legs, which she isn't afraid to set off in jaunty A-line skirts that end just below the knee. A devoted wife and mother of two sons and a daughter, Ramerman is ferociously committed to justice for women in the Catholic Church.

From its earliest beginnings, Ramerman's ministry was broad and daring. That first year, Callan asked her if she would preach at the Mass on Mother's Day. Stunned at first, she obliged. But when he asked her to preach again, she panicked. "Every thought I've ever had I just put into that sermon," she says. "I couldn't possibly do it again." As it turns out, she could. The years went by, and Ramerman kept on preaching. She also visited the sick, prepared people for marriage and baptism, helped administer the parish, and worked to launch an explosion of outreach ministries—to the poor, the homeless, the sick, the dying, the drug-and alcohol-addicted, ex-offenders, and abused women—which grew to include seven hundred volunteers. Building on its early ministry to people dying of AIDS, Ramerman helped lead the way in Corpus Christi's creation of a gay and lesbian ministry.

A big step forward for Corpus Christi and Ramerman came in 1988. At Callan's request, Father Enrique Cadena, who had been active in Mexico's small base communities and in Corpus Christi's Mexico mission to the poor, joined Corpus Christi parish. He was committed to conveying to the growing congregation what Ramerman described as "the holiness of the people." In keeping with that, he asked Mary Ramerman one night if she would like to hold the chalice with the sacred wine while he held the bread during the service of Mass. She agreed. So at an otherwise ordinary evening Mass at Corpus Christi Church, at the moment of Eucharistic consecration, it was not Father Enrique but Mary Ramerman who raised the chalice high up above her head.

Denise Donato remembers the moment well. She had just joined Corpus Christi, after leaving her parish of many years, deeply wounded by her priest's failure to validate her call to the priesthood. "The tears just would not stop the whole service," she recalls. "Now Mary never said the words of consecration or the Eucharistic prayer, the words reserved for the priest, but seeing her and knowing she had a presence and a voice was very, very powerful." Ramerman's impression of the reaction of the people at that moment was that they were "absolutely stunned." Some, she confesses, were "extremely angry." Indeed, confirms Callan in his book, some left the parish because of that action. But "others were overjoyed to see a woman have a liturgical role at the Lord's table."

As Ramerman's ministry grew and she was promoted to associate pastor, it became clear to her parishioners that her clothes were all wrong. An eighty-year-old parishioner, Florence Bartels, commended Ramerman on the job she was doing but said it wasn't fitting for her to "celebrate the Eucharist in a paisley dress." Bartels recommended Ramerman buy a white dress, which she did. But that proved to be only a temporary solution. In time, Corpus Christi's liturgy committee took matters into their own hands. Because Ramerman held the same post as Father Enrique, the committee decided her clothing needed to reflect that. So they authorized Ramerman to wear priests' garb—a long, white, linen tunic with sleeves, called an alb, and a stole. Actually, what they gave her was a purple half-stole, reflecting her nonordained status. Still, it was a provocative act; the stole is forbidden attire for anyone but the ordained. Ramerman stepped onto the altar in that clothing for the first time in 1993.

Soon Ramerman was participating in all aspects of the Mass, from the opening prayers to the peace prayer to the final blessing, as well as in funeral services, weddings, and baptisms. She also began to lead Communion services, distributing already consecrated hosts, when the priests were out of town. Then one night, a woman who had been physically abused by a man asked at an evening reconciliation service if she could confess to Ramerman rather than a priest because that would be easier for her. Callan agreed. In time, others began to ask. Though Ramerman only listened and left it to the priests to absolve, her line grew longer and longer, until it was the longest of all. "Mary couldn't get away because so many people, both men and women, wanted her to listen to their confessions," wrote Callan. "Parishioners kept calling Mary to do more and more priestly things."

Ramerman stresses that none of these changes were made blithely or quickly, and most were made by committee. For example, after Ramerman began to hold the chalice, the congregation launched a yearlong educational program, inviting professionals from Saint Bernard's Seminary to come and talk about the history of the Church and the Eucharist. As a result, the parishioners agreed that they did want women ministers on the altar holding the wine.

Over the years, Corpus Christi Parish grew from a failing enterprise with barely three hundred parishioners to a vibrant congregation of more than three thousand, in wholehearted support of its inclusive ministry. The gay and lesbian ministry expanded to the blessing of gay unions in formal commitment ceremonies, as well as a preparation program for gay couples. Corpus Christi extended Vatican II's commitment to a new ecumenism by inviting all to the Communion table. And the women's ministries grew exponentially. Besides Ramerman, women like Sister Margie Henninger preached. Others held Communion services when the priests were unavailable, including a nun who was a college chaplain; two Hispanic women who held services in Spanish; and Denise Donato, married mother of three, who was hired as family minister in 1995.

An Enemies List Grows

Despite the jubilation in the pews, Corpus Christi was attracting plenty of enemies, too. Parishioners were sure there were spies about, shutters clicking, recorders taping. For many years, the bishop of Rochester, Matthew Clark, himself a Vatican II bishop and a longtime supporter of expanding women's roles in the Church and of lay-clerical collaboration, participated in and even encouraged Corpus Christi's work, reports Ramerman. But as the spirit of Vatican II was snuffed out by the orthodoxy of John Paul II, Clark pulled back. The reports from Corpus Christi's detractors didn't help.

The situation reached a crisis in 1998. On orders from the Vatican Congregation for the Doctrine of the Faith, Bishop Clark was forced to take action. As Callan reported it, a letter dated July 25 from Cardinal Joseph Ratzinger landed on Clark's desk. That letter, which Callan was never allowed to see, ordered Clark to remove Callan from Corpus Christi and put a "trustworthy priest" in his place "to bring things in line with Vatican policies." Clark delegated the task of telling Callan to Father Joseph Hart, the newly named vicar general of the Rochester diocese. Callan was advised of his three transgressions: open Communion, blessing gay and lesbian unions, and allowing women on the altar during Mass.

Callan didn't even attempt to dress up the truth. He told his parishioners that he was being banned for being inclusive. Stunned and enraged, his parishioners took action. Over 1,400 of them packed the church on the last night before his transfer. "You can't hold back the spring," read the banner they strung across the front of the church. They sang. They spoke. They organized demonstrations and letter-writing campaigns to Bishop Clark, and developed a "Statement of Faith" on the three inclusive policies, overwhelmingly approved by a vote of 3,022 to 7.

Despite the uproar, Callan was out, on a temporary transfer to a small parish in Elmira, New York. Mary Ramerman stepped into the breach, conducting Communion services. In one of her last Sunday homilies, she said: "If the Church stands up and preaches God's love and at the same time devalues women by not allowing them to be on the altar or to preach from the pulpit or to wear the appropriate liturgical garb or whatever the Church might do to keep women away, then it makes a mockery of the Gospel. In order to be the light of the world, we first have to be a light in the Church. We first have to look at our own discrimination."

Her intransigence did not go unrewarded. Ramerman was ordered by the bishop to take off the alb and the stole and step off the altar. "I was an associate pastor there for fifteen years," says Ramerman. "Then all of a sudden I was told, 'Every time we have Mass, you're not to go anywhere near the altar.'" It was an order Ramerman could not abide. "What message would that be giving to the men and women who come in the Church?" she says. She refused to step down. On October 15, 1998, she was fired.

Called to Serve

While many Catholic women have clear and frustrated longings to be priests, Mary Ramerman was not one of them. "I was very happy in my role at Corpus Christi," she says of her days before the decimation of the parish. "I was able to do everything I wanted to do. I was able to minister in many different ways. It really wasn't until I had to make a choice between continuing to do that ministry or be subservient to a Church teaching that I couldn't believe in that I began to see myself as a priest."

After Ramerman's dismissal, other women in the parish took up the mantle of women's equality. They wore purple stoles, a symbol of women's ordination, and stood in her place on the altar. But the sweep of ministers from Corpus Christi proceeded in earnest. Father Enrique resigned. As Callan reports, he himself refused to sign a statement saying that he would "never again allow a woman to have a prominent role like Mary's on the altar, never again invite everyone to the Communion table, and never again bless same-sex unions." For that, he was fired from the diocese, too. Unlike Ramerman, he was offered a monthly stipend, which he refused because the same offer had not been made to Ramerman.

Denise Donato, Corpus Christi's family minister, remembers the devastation all of them felt. "There was one weekend in particular that was very difficult, the weekend after Mary was fired," she told me. "People were just in mourning. I remember being at the [Saturday] five o'clock Mass sobbing, and the people around me were sobbing. I couldn't even look up at the altar." On Sunday, Donato stationed herself outside the church from 7:00 in the morning until 1:30 in the afternoon to offer comfort and hope to those who couldn't bear to stay inside. She was never out there alone for more than a few minutes. Then, on December 14, just after his arrival, the new priest assigned to Corpus Christi—whose job it was to bring the parish into line with the Vatican's policies—fired six more people who had devoted decades of their lives to the parish and its ministries. "These six people, most of whom had families to support, were fired just ten days before Christmas and, like Mary, were given no severance pay or benefits," wrote Callan. One of them was Denise Donato.

Mary Ramerman put a public face on the decimation of Corpus Christi. She wore that face with courage, even defiance. She talked to reporters. She took to heart Sister Joan Chittister's legendary advice, spoken at the Women's Ordination Worldwide Conference in Dublin. "She gave a phenomenal talk there," Ramerman recalls. "One of the things she shared [was] 'If you're going to stay in the Church, stay loud. If you're going to leave, leave loud.'" Ramerman had chosen to do both—*loud*.

The parishioners needed a place to be together to come to terms with what had happened to them, to decide where they belonged. Ramerman became the leader of their "Church in exodus." She found a place for the expelled community to meet for services on Tuesday nights, at the Downtown United Presbyterian Church. Soon, they had offers of other spaces, too, for weekday and weekend services, at Immanual Baptist Church, the Salem United Church of Christ, and the Hochstein Performance Hall. The hall is a portentous place, Ramerman and everyone at Corpus Christi will tell you, because it used to be the Central Presbyterian Church where suffragist Susan B. Anthony and abolitionist and suffragist sympathizer Frederick Douglass worshipped and where their funeral services were held.

Hundreds of people came to the services that Ramerman led. She preached, consoled, and officiated, but still never consecrated the Eucharist. By the beginning of 1999, this pilgrim parish had formed its own community, which they called Spiritus Christi. Callan participated in one of their services, which got him expelled from the priesthood. But it also convinced him that Spiritus Christi parish was where he belonged. Father Enrique returned, too.

To the Rochester diocese, what was happening was a disaster. Officials characterized the emergence of Spiritus Christi as a schism. They announced that all members of Spiritus Christi by their actions "had chosen to excommunicate themselves." But many parishioners rejected that conclusion. "None of us really believes in excommunication," says Ramerman. "I don't know what excommunication is. It's totally a foreign concept to the Scripture, to Jesus."

Ramerman had no intention then—nor does she now—of leaving her Catholicism behind. She considers herself Catholic and loves the tradition. "I love the liturgy," she says. "I love the earthiness of it—Catholics have incense, oil, water, candles. It's a very tangible religion. You can touch it; you can feel it; you can smell it. I love the prayers and the rituals, its history of saints and sinners. . . . I love the wrestling with theology, with the concept of personal conscience. . . . It keeps people asking questions, wrestling with issues. I think it's a very important part of faith, to be able to wrestle with God."

As a Catholic woman, Ramerman considers herself "part of the woman's story of Catholicism," a very long story of women's history serving the Church, "as nuns, as pastoral assistants, as ministers to the homeless, in food programs, cleaning the altar cloths, providing the music," she says. "This Church would be *nowhere* if there had not been women in it for two thousand years." To Ramerman, it is critical that "as women, we have a voice in our own Church. I don't accept being thrown out for asking to have a voice. . . . I'm just not going to accept that."

Spiritus Christi soon grew to a congregation of nearly 1,500 people. There were baptisms, communions, confirmations, weddings, and funerals, a busy program of religious education, and many missions to the poor and disenfranchised. But there were still just two priests, both men. The parishioners wanted that to change. They wanted Mary Ramerman to become a priest. "A lot of people said, 'Mary, just do it. Just celebrate the Eucharist,'" reports Ramerman. "You're our priest. You're leading the services. Why don't you just do it?" But she could not. "I didn't want to become a church where anything goes, where whoever feels like doing something does it. . . . I think being called is important, having prepared ministers . . . is important." She said she found she was "unwilling to be ordained until we had found the right path."

Forging a Path to Ordination

To determine how exactly to ordain Ramerman, the parishioners on Spiritus Christi's women's ordination committee began to meet regularly with Ramerman. They reviewed the components of the Catholic sacrament of ordination. In 2000, in partnership with the Women's Ordination Conference, they organized an international meeting in Rochester to further discuss the possibilities, to which some 250 people active in the global women's ordination movement came. That global movement had been buoyed by WOC's publicizing the discovery of a woman officially ordained a priest by a Catholic bishop in good standing, in 1970. Ludmila Javorova served as a priest in the underground church in Communist-occupied Czechoslovakia until communism fell, when her services were no longer required and the Vatican repudiated her ordination. In the late 1990s, WOC brought Javorova to the United States for a private tour.

The committee members also met with women ordained in other denominations. Finally, they identified three aspects of ordination that were important to them. One was having representation by the Spiritus Christi community, which was calling Ramerman to be its priest. Another was having an interfaith component, with representatives from other traditions confirming her call. And another—by far the greatest challenge—was having a bishop to do the ordaining, the traditional laying of hands, signifying the long tradition of Catholic Church history.

"Many people told me: 'You don't need a bishop,'" reports Ramerman, "just have the community do it." But that didn't work for her, either. She wanted some way to acknowledge the Catholic tradition, which requires a bishop to ordain a new priest. But finding a bishop to ordain her, in the face of the Church's adamant prohibition against even discussing women's ordination, was daunting. While there are Catholic bishops in the United States in sympathy with the cause, and many abroad, none was willing to come forward and take the risk.

Then Ramerman heard about Bishop Peter Hickman. She heard about him from another Roman Catholic minister, the Reverend Kathy Mc-

Carthy. Like Ramerman, McCarthy had come to a point in her life where she was ready to be ordained a priest. Her solution to the vexing problem of how to get this done was to be ordained in the Old Catholic Church; her ordination by Hickman had been scheduled. At first, Ramerman rejected the idea outright, fearing it was another "flaky" route to ordination. But as she learned more, her "prejudice" against the Old Catholics began to melt away.

The Old Catholic Church was for centuries part of the Roman Catholic Church. That partnership endured until 1870, when the doctrine of infallibility—which holds that the pope cannot err in matters of faith teaching—was adopted at the First Vatican Council. That new doctrine caused a great furor within the Catholic Church, one casualty of which was the break with Rome by those who now call themselves the Old Catholic Church. However, because of its history, the Old Catholic Church remains Catholic (that is, it is not among the churches that broke with Rome in the sixteenth-century Protestant Reformation). As such, it shares the Roman Catholic Church's so-called line of "apostolic succession." Its bishops, like Roman Catholic bishops, are descendants of the bishops who took up the mantle of the Christian Church after the death of the apostles, which means that Old Catholic bishops like Hickman provide a valid line of apostolic succession—an essential element of ordination. Film actor and director Mel Gibson (*The Passion of the Christ*) is often described as an Old Catholic, but he does not seem to be affiliated with the institutional Old Catholic Church. A self-described staunchly conservative Catholic who rejects the teachings of Vatican II—for example, the English Mass—Gibson reportedly is building his own church for his own group, the Holy Family, on sixteen acres near Malibu, California.

Slender, graceful, with thick red hair and a beard, Hickman is a Kenneth Branagh look-alike and drop-dead gorgeous. He's the first to say he is not typical of Old Catholic bishops. While some other Old Catholic bishops favor women's ordination and have ordained women in Europe, where the Old Catholic Church has more members, most are far more conservative than Hickman and his community, Saint Matthew Church in Orange, California. Ramerman contacted Hickman, who invited her to visit his parish to attend a retreat for his priests. She did. In turn, he traveled to Rochester for a grilling by the "Spiritus Christi community of rebels and very challenging people," as Ramerman describes them. "He was fantastic," she says. "Very solid in theology, very pastoral with people." He survived the grilling and went back home. Spiritus Christi congregants accepted him. They voted officially to make Ramerman a priest. Hickman would preside. The ordination was on.

When word got out, all hell broke loose. The Rochester diocese ran an official pronouncement in its newspaper, the *Catholic Courier*. Vicar General Father Joseph Hart decried the ordination, saying that "one cannot honor this terrible, disruptive breach by attending this ceremony as if nothing had gone on, as if this is someone from the community that never had any relationship with us." He ordered away from the ordination and from participation in sacraments at Spiritus Christi all clerics, religious, and anyone "in a teaching position in the Church," for they would "risk scandalizing the Catholic faithful" and "give the false impression that Spiritus Christi is a Catholic parish." In local parishes, many priests also cautioned their parishioners not to attend the ordination.

Ramerman says she received a letter from Bishop Clark, in which he blamed her "personally for scandal in the Church," held her "responsible for the pain of people in this diocese," and prevailed upon her "to abandon your leadership role at Spiritus Christi." Ramerman was "stunned" by his reaction. "I just wondered: Does he really not realize that this pain has been going on for a long time? There is tremendous pain in the Catholic Church, tremendous pain with women and discrimination. I can't even give myself the honor of being solely responsible for it. I'm not that powerful."

It's worth noting that while Ramerman was declared a schismatic and totally banished from the Catholic fold for her association with the Old Catholic Church, Mel Gibson was welcomed by the Catholic hierarchy with open arms when his film came out. At the time, a Vatican official told the *National Catholic Reporter* that "while Gibson may be a bit idiosyncratic theologically, 'his heart is in the right place.'" Mary Ramerman's struggle for equality in the Church was condemned as a scandal that "wounded" the Church, but Gibson's bold defection from the Church to start his own was overlooked as a mere foible.

While there was an outpouring of joy over the upcoming ordination, there was also great consternation. Ramerman admits that "many people were angry with me. They said: 'It's not time yet. It's not time for women to be ordained. It's not time to stand up against this.' But that's what we said about many things, isn't it?" To Ramerman, two thousand years was more than enough time to wait.

Oh Happy Day

It's nearly 10:00 A.M. The crowd at the Eastman Theatre is overflowing now, standing room only. The stage brims with carnations and lilies. The gospel-inspired chorus dazzles, all in white. The orchestra has settled in. At 10:00 A.M., bells ring. Three thousand people rise to their feet. Mary Ramerman's ordination begins.

The atmosphere is electric, full of music, dance, jubilation, and song. Slowly at first, sweetly, come the strains of "We Are Standing on Holy Ground." Then we are blasted out of our seats by the bold and insistent lyrics of "Come and Rejoice, Sing a New Song." The aisles spring to life with white and pink ribbons waved by dancers from six-foot poles. Children sail across the stage bearing gifts of poster art, like the scales of justice, equally balanced. Father Jim Callan practically jumps out of his seat onstage, tapping his toes and beaming.

The ceremony commences. Ramerman is radiant. So is Hickman. "Mary, as I stand before you this day, I am deeply aware that there is another bishop, a resident of the Catholic diocese of Rochester, who ought to be here participating with you in this great moment of your ordination," he booms. "Nonetheless, the fact that we are all gathered here today in this great and holy city of Rochester, the city of Frederick Douglass and Susan B. Anthony, the city that has so oftentimes served as the conscience of this nation, bears witness to the fact that the . . . will of Spirit will prevail over the hardened hearts of men, that the justice of God shall overcome all the injustices of humanity."

Hickman brings with him a holy rage: "Enslavement to a tradition that does not reflect the Gospel of Jesus is enslavement to the path of tyranny," he bellows. "Women cannot be abused for the sake of tradition. . . . Every injustice against women is an injustice against the whole Church."

Ramerman's husband reads the first reading. Her son, John, the second. The Spiritus Christi community calls forth the candidate. "Bishop Peter," begins each person who arrives on the stage, "on behalf of the children . . . our families . . . gay, lesbian . . . all who practice peace and justice . . . in prison and in jail . . . suffering from alcohol and drug addiction . . . women who endure poverty, discrimination, violence, and oppression . . . we call on Mary to be our priest." For the laying on of hands, besides Bishop Hickman, more than sixty spiritual leaders and Church reformers from around the world, as well as more members of Spiritus Christi, come up to the stage where Ramerman kneels, each in turn stroking her hair, holding her face, staring intently into her eyes.

While Ramerman lay prostrate, signifying her surrender, the parishioners call out the litany of the saints. "To all you holy men and women, pray for us," they say, calling on Wisdom-Sophia and Mary of Magdala; on Sadie Wilson, who taught Ramerman about Jesus; and on Florence Bartels, who urged her to be a priest. Ramerman's best friend vests her with priestly clothes, a beautiful white chasuble, a green-striped stole. When it comes to the consecration, all of us speak the words. All of us receive Communion.

Throughout the ceremony, the music soars, from "Claire de Lune" to "Oh Happy Day," from "How Can I Keep from Singing" to "Faith Has Brought Us Here." The audience cheers, sings, laughs, claps, prays, and cries. I cry, a lot. "God is doing something awesome in our midst," said the parishioner who gave the welcome. Indeed, there was something awesome about being there, an amazing force unleashed in that ceremony, in that unequivocal act. There is no underestimating the power of seeing divinity—in this case, a woman priest as a stand-in for Jesus—represented in your own form, and no underestimating the pain caused by a beloved Church that refuses to allow women to represent that divinity.

At the end, Ramerman is asked to give her first blessing as a priest. It is the first time during the ceremony that she speaks. She comes to the microphone. She smiles. She thanks everyone for coming. She pledges to be the best priest she can be. She says she'll think about how she'd like to be addressed, since people keep asking her. But it is her very first words that move me the most. "This is way cool," she says to the multitudes gathered in her name. And everybody claps.

At the press conference after the ordination, Ramerman is asked the hard question: How can you make change in the Roman Catholic Church from the outside? "We're doing it," she says, as thousands leave the theater carrying an image of Catholic divinity that they had never seen before. "As my mother used to say, 'The proof is in the pudding.'"

The Men at the Top

Bishop Peter Hickman turns out to be an accidental activist. Which I guess is why he sometimes looked surprised up on that stage. "This is not my brainstorm, my bright idea," he tells me, smiling. "It's not like I'm driving the train. The wave of the Spirit's happening, and I'm just catching the wave."

He fell into this role, he admits. A long time ago, a woman member of his small faith community, Saint Matthew's Church, asked what he thought about women's ordination. He had to confess he hadn't thought about it much at all. But then he did, concluding that he was indeed in favor of it. So when a woman asked him to ordain her, he was ready. Then came Mary Ramerman. Later, others will come. With his bishop's miter and his staff, Hickman, for Catholic women, has become a symbol of connection from past to present. And he's okay with remaining that symbolic missing link for now. He'll keep doing it because "it's a matter of conscience; it's a matter of justice; it's the right thing to do."

On the other side of the divide is Bishop Matthew Clark. He has been an advocate for women in the Catholic Church for decades. He served on the committee that listened to the testimony of tens of thousands of Catholic women for the ill-fated U.S. bishops' pastoral on women. In 1993, after that pastoral failed to pass, he held a diocesan-wide synod in Rochester, out of which came recommendations for women's ordination and married priests, which he relayed to the Vatican. And in 2004, he was the only bishop who came to Boston College's "Envisioning the Church Women

Want" conference, to serve on a panel entitled: "When the Bishops Listened to Women: The Women's Pastoral Twelve Years Later."

In that workshop, Bishop Clark shared the thinking of women he'd spoken to about the Church they want, saying that their thinking "is very much my own." He talked about a liturgy he'd recently attended where some thirty people played leadership roles, men and women, the pastoral administrator and the priest from Sri Lanka, where everyone was invited to take an active part. It was not a big city parish, he said, and no one was telling them that this or that was not allowed. He said the kiss of peace at that service was "close to the most religious experience I've had in years."

I approach him after his Boston College workshop. I tell him about the book I'm writing, that I've heard so much about him, and that I'd like to talk with him about Mary Ramerman's ordination. I can't tell how he feels about me or what I'm doing. He gives me his number and the name of his secretary. He advises me to tell her that he's expecting my call.

I call him twice before we finally connect. He cautions me that he hasn't decided if he'll talk to me or not, but he'll hear me out. I tell him exactly who I am, who I've spoken to, what I'm doing, and what I'd like to ask him. I say that his name has come up many times on my travels as a bishop admired and respected by many Church reformers. Because of his involvement with the role of women in the Church, I say I found it ironic that he of all bishops was the one faced with the crisis at Corpus Christi. I wonder what that was like for him.

When I finish, he says, in a low, even voice: "I think I'm going to pass." My heart falls, but I'm not surprised. Then he says, softly: "I hope you don't take it unkindly. It is for all kinds of reasons right now." That does surprise me. I believe I detect a sadness in his voice, a tone of defeat, which I also thought I'd read on his face at the Boston College event. I had been told he'd been ill, that what happened at Corpus Christi had been very painful for him. A feeling of sympathy rose up in me. "I wish you good luck, Bishop Clark," I heard myself say. "I wish you luck, too," I think he said back. As I hung up the phone, a deep sense of hurt and loss washed over me. I felt a wall rise up between us, a wall he had clearly been forbidden to scale. I stood at the foot of my side of that wall, helpless.

Later, I looked over my list of questions for Bishop Clark, especially the last one: "How do bishops and priests who believe that excluding Catholic women from sacramental ministry is wrong, sinful even, blatantly discriminatory, justify their silence and their accommodation to this exclusion?" It is the question I most regret never having had a chance to ask.

But Bishop Clark represents the past. Mary Ramerman is the future. While it may not look the way anyone expected it to, the ordination of women in the Catholic tradition has begun.

14

WOMEN'S ORDINATION: CALLING THE QUESTION

Within eight months of Mary Ramerman's ordination, on June 29, 2002, seven more women were ordained Catholic priests. The ceremony took place on a cruise ship, the *M.S. Passau*, on the River Danube, just outside of Passau, Germany, before an audience of several hundred. Having the ordinations in international waters was intended to foil any attempt by a Roman Catholic bishop to claim jurisdiction and interfere. The event was shrouded in secrecy—the location, the identity of the bishops, the candidates.

One of the women ordained was Austrian Christine Mayr-Lumetzberger. Once a nun who later married, Mayr-Lumetzberger, forty-seven, developed a three-year ministerial training program for women who want to be priests, which was approved in 1998 by the General Assembly of We Are Church-Austria, a member of International Movement We Are Church. She was among the first women to complete it, as did five of the other women ordained on the Danube. All had theological training.

As a brilliant sun rose over the ship's deck, the service began. Presiding was the controversial Argentinean Bishop Romulo Antonio Braschi. An ordained former Roman Catholic priest and bishop, Braschi founded and heads the independent Catholic Apostolic Church of Jesus the King, is married, and has been excommunicated by the Vatican. The complications in his status—and that of the attending former Roman Catholic priest, who Braschi ordained a bishop—contributed to the refusal of major European progressive Church reform groups, including We Are Church-Austria, to support the ordinations. That was deeply disappointing to the women ordained. But other groups did support them, including the U.S.-based Women's Ordination Conference. And the women ordained were satisfied that Braschi, with his roots in the Roman Catholic Church, provided the "apostolic succession" required by the Roman Catholic ordination rite.

Upon learning of the consummated ordinations, the Vatican immediately retaliated. In dramatic contrast to its coddling of priest child sex abusers, the Vatican threatened to excommunicate the new women priests if they failed to apologize and renounce their ordinations in one month, by July 22. Ironically, the Vatican either did not realize or chose to ignore that July 22 is also the feast day of Mary of Magdala, today marked by hundreds of women-led celebrations that attract thousands of Catholics worldwide. The new priests appealed the decision, but their appeal was denied. The Congregation for the Doctrine of the Faith formally excommunicated them, forbidding them from celebrating or receiving the sacraments and publicly chastising them for having "wounded" the Church.

These newly ordained women follow the tradition of the controversial French "worker priests," who lived and held ordinary jobs among the people. Mayr-Lumetzberger and another of the new priests teach while conducting their ministries; two others, including a former nun who was expelled from her religious order, work with the elderly and the terminally ill. The only American woman, Austrian-born Dagmar Braun Celeste—ex-wife of Ohio's former Governor Richard Celeste—visits parishes and Catholics in her diocese, by invitation; one priest asked her to meet with his aged mother, who, he told Celeste, had waited her whole life to see a woman priest.

As for Ida Raming, less than a year after her ordination, in the spring of 2003, this septuagenarian set off on a road trip. Grabbing her longtime companion, another newly ordained priest, Iris Müller, Raming left Germany to embark on a five-week, twelve-city speaking tour that would show America what a Catholic woman priest looked like.

Raming's Road to the River

Ida Raming was born in a rural community in Germany. As a child, she served Mass in the woods, at an altar that was actually a tree, using a chalice that was actually a glass. Her first love was always religion, which led to Raming becoming one of the first women in Germany to enroll in a graduate theology program. It didn't take much time to see that pursuing theology in the Roman Catholic tradition was a dead end. "During my studies, it weighed on me dreadfully that I failed to find a spiritual or professional niche for myself," Raming wrote. "I suffered greatly in this spiritual homelessness."

Then, in 1961, Raming met Iris Müller. Unlike Raming, who had been discouraged in her studies from criticizing the Church, Müller felt free to criticize it openly and often. Müller believed that ordination was a clear "prerequisite for justice for women in the Catholic Church," wrote Raming. In 1963, Raming and Müller, then doctoral students at the University of Munster in Munich, drew up a theological defense for the ordination of women in the Catholic Church. German lawyer Gertrud Heinzelmann included that piece with other women's contributions in a pioneering publication of women's voices called *We Shall Keep Quiet No More! Women Speak to the Second Vatican Council*. When Heinzelmann could not find a publisher, she published it herself, then gave it out at the council meetings.

In the late sixties, Raming dived into her groundbreaking dissertation. Her thesis was that the canon law restricting ordination to baptized men was indefensible, unjust, and illegal because it effectively authorized two classes of baptism. Published in 1976, *The Exclusion of Women from Priestly Office—A God-Given Tradition or Anti-Woman Discrimination?* won her no friends in high places. Raming was promptly closed out of university-level religion teaching jobs in Germany. She found other teaching work but never

left the Church, to the enormous consternation, decades later, of her enemies and some of her friends.

It's April 28, 2003, and Ida Raming emerges into the lobby of the Beacon
Hotel in Manhattan. She descends a few steps to the waiting area where I
am standing and takes my hand. She leans in to tell me who she is, as if I
needed an introduction. "I'm Ida [pronounced *Eda*] Raming," she says in
slow, labored English, tightly squeezing my palm.

Raming is heavyset and broad-shouldered. She wears a pastel blue
pantsuit, with black supportive shoes. "Her psoriasis is so bad her feet bleed
when she walks," says Rea Howarth of Catholics Speak Out, cosponsor with
the Women's Ordination Conference of Raming's U.S. tour. The observation strikes me as oddly biblical. Raming takes charge, in a quiet but insistent way, always with a soft smile that belies the authority in her nature. This
is a woman who has taken stands, whose whole life right now is a stand. "It's
too noisy here," she says of the Time Café, where I take her for our interview.
"This chair is uncomfortable," she says, so we move. "I don't want anything
spicy in my food," she says. "I eat very bland." I make a recommendation that
she orders and enjoys, which inexplicably makes me proud.

Raming tells me that the tour is going well. It has brought out hundreds
of people, from nuns to laywomen to priests, in cities from Palo Alto to St.
Louis, Minneapolis to Chicago, Washington, D.C. to Boston. Raming's
New York City appearances are scheduled for tomorrow. One is to be at
Fordham University, a Catholic school run by the Jesuits, where you might
expect some opposition to the appearance of an excommunicated Catholic
woman priest. The other is to be at Union Theological Seminary, a nondenominational school known for fighting discrimination, including religious
discrimination, where you would expect a warm welcome.

As it turns out, the small but enthusiastic event at Fordham ruffles no
administrative or hierarchical feathers. But Raming's visit to Union is another story. Had it not been for a feisty, young theology student named
Leslie Kretzu, there might not have been any Raming event. Certainly, there
would have been no Mass.

A Little About Leslie

It's ninety degrees and climbing in Manhattan. Leslie Kretzu and I meet for lunch in a noisy restaurant just below Canal Street to talk about Raming's visit four months ago, in April, to Union. Kretzu is coming from a nearby studio where she is working with her fiancé on a feature-length documentary called *Sweat*. At twenty-nine, she has just graduated from Union, with a master's in theology and a specialty in ethics. On this day, her thick, dark hair is pulled back off her face. She has coal black eyes and a big, wide smile. Her shapely body is snuggled into a pair of comfortably worn jeans, and she wears a hot pink sleeveless top, out of which peek black bra straps.

Kretzu grew up a traditional Catholic, a product of Catholic elementary school, high school, and college. She began during those years to get a glimpse of the notion of Catholic social justice, so when she graduated, she joined the Jesuit Volunteer Corps. She moved to San Francisco and became ensconced in a lively and diverse Catholic faith community of people working for social justice. One night she joined an audience of eleven in a little chapel in Berkeley to hear a speaker from the National Labor Committee talk about sweatshops—the intolerable working conditions, the high cost of goods sold, the unspeakably low wages. That presentation set Kretzu on fire; she was determined to join the fight for sweatshop workers' rights.

While Kretzu was still in California, one of her friends told a young man in New York about Kretzu's new passion. The young man, James Keady, already was a seasoned advocate for the rights of sweatshop workers. By then, he had sued his alma mater, St. John's College in Queens, New York, where he'd been studying theology and working as an assistant athletic coach. Claiming he had lost his job for refusing to wear Nike apparel to practice and training events—a requirement of the university's multimillion-dollar Nike agreement—Keady sued the school and Nike for $11 million, a case he would ultimately lose. He found that requirement completely untenable, given what he characterized as Nike's deplorable labor practices and in light of Catholic social teaching in defense of workers' rights. Keady contacted Kretzu online and told her about himself and his work. It made a great impression. "Here's this twenty-six-year-old Catholic kid telling a Catholic university they're not being Catholic enough," says Kretzu, laughing. "I was in love with him before I met him."

Kretzu soon returned to New York, met Keady, and confirmed that it was love. About a year later, in 2000, she moved with Keady to Indonesia, maxing out their credit cards to make their film. They settled in an industrial zone with an open sewer system where Nike's factory workers live and moved into a "nine-by-nine cement box" that was stifling in the one-hundred-degree heat. The two failed to get jobs at Nike but determined to try to live on Nike's wages of $1.25 a day. They also interviewed Nike workers trying to do the same. Their conclusion: It is impossible. They found parents "working for the largest sporting goods company in the world who had to send their children to their home villages because they could not afford to keep them," says Kretzu.

After she returned from Indonesia, a friend suggested Kretzu go to theology school. It was something she had never before considered, even though she saw herself as a human rights advocate with a faith-based foundation, who was given to curling up with liberation theology texts instead of novels. The more she thought about it, the better the idea sounded. Kretzu had no idea where to go, so she Googled "theology and New York" and up came Union. She found out that Dr. James Cone, a highly regarded black liberation theologian for whom she had great respect, taught there. She expected that studying theology "would help me to think more critically, solidify the foundation" she needed for her work. Beyond that, she just felt, "I need to go."

Kretzu was in her final year at Union when one of her professors, Sister Janet Walton (with whom Kretzu participated in a women's liturgy group), learned about Ida Raming's speaking tour. Walton suggested that Kretzu organize a visit by Raming to Union. As a member of Union's social action caucus, Kretzu could book the chapel, print and post flyers, and hold the event. Even though she was incredibly busy—finishing school, speaking on behalf of the new nonprofit she and Keady had founded called Educating for Justice, and planning her wedding—she was strongly in favor of women's ordination and so said, "Why not?" She contacted an organizer of the tour,

Rea Howarth of Catholics Speak Out, who was pleased that Raming was invited to give a lecture. But then Howarth asked: "Would you mind if we did a liturgy?"

The Sides Line Up

Kretzu realized that a Mass celebrated by an excommunicated female priest, even at Union, could be controversial, so she began by asking the Catholic community what they thought. She put together a list of nearly fifty people, including about thirty students (out of Union's two hundred or so degree students), faculty members, and staff. She sent out an e-mail, along with an article about Raming. She aimed for a neutral tone, especially regarding the Mass, which actually she felt ambivalent about. "I wasn't like a hundred percent gung ho because there was so much else I was doing." She asked the students to let her know what they thought.

Then she waited. And waited. "Nobody responds after a week. This is the Catholic Church right here," says Kretzu. "Nobody says a word about what's going on. . . . It's why we're in the situation we're in." Later, she amends that slightly. "There's like one or two people looking to do things in a more just way . . . and you have the majority being silent." She waited one more week, then sent out another e-mail, stronger in tone than the first. "I'm writing this e-mail to let you know that if we have the liturgy, it's going to be a groundbreaking event. Your comments would be really appreciated."

Finally, the replies began to come in. Her first e-mail came from Professor Mary Boys, a highly respected authority on Christian-Jewish relations, one of the experts who publicly criticized the anti-Semitic content of Mel Gibson's film *The Passion of the Christ*. As Kretzu reports, Boys "is about 110 percent against" the liturgy. Among her reasons, according to Kretzu, were that it would be seen by many people as an unnecessary act of defiance and would reduce the Eucharist to a political act.

"I almost fell off my chair," says Kretzu. "And now I'm angry. Now I really want to do the liturgy." That feeling only deepened as other Catholic PhD students wrote back saying it was "a bad idea." Among them was Luis Ramos. "He used all this totally flowery language," reports Kretzu, along the lines of "the oppressed peoples of the world are going to have their time when they will be able to rise up." Kretzu shot back: "Luis, would you mind telling me when this time's going to come?"

Kretzu took Boys's e-mail, which had been sent to everyone on the Catholic list, and responded point by point. "I said we're not the ones making the Eucharist a political act. We are the ones struggling for equality." And, Kretzu said, "every act of social change is deemed by at least one group of people to be an 'unnecessary act of defiance.'"

Professor Boys was at first unwilling to talk to me about what happened at Union, but then she agreed to speak. She called the small Catholic community there "fragile" and worried about baring any more Catholic "wounds" in public. She admitted to long supporting women's ordination but felt conflicted about Raming's celebrating Mass. She did see it as an unnecessary act of defiance but noted that she is not against such acts "in themselves." Her main contention was that Raming's ministry was not "something that came from the inside," from Union. "I felt like we were being used a little bit." She also felt that Raming's celebrating Mass would be a politicization of the Eucharist, again, because Raming was a stranger to the Union community. While Raming sees herself as a minister to the worldwide Church reform community, Boys sees her as a rootless itinerant minister who travels about presiding at the Eucharist. "That's not what priesthood is all about, either," Boys told me. "I guess it was a kind of an exploitation, at some level. I don't mean in any way to attack her integrity. . . . It's a matter, maybe, of a different reading of the Eucharist."

Finally, Boys admitted to not wanting the event, which had already engendered turmoil among Catholics at Union, to cause "the cardinal to swoop in and make it uglier." More specifically, she was quoted in an article in Union's student newspaper as saying that "inviting Raming to preside at Eucharist may be regarded by people in power as highly inflammatory—and many of us who do work within the structures of the institution could have that work inhibited by our association with the event."

That argument incensed Kretzu the most, even though she understood that as a woman who planned to work outside of the institutional Church structure, she had much less to lose. "They are women . . . who have some power, but not that much power in the Catholic Church," Kretzu said. "They're trying to hold on to—what? Their lack of power?"

Kate Ott is one of the graduate students who opposed Raming's celebrating a Mass. She explained to me that she had concerns about Raming's motives. "Ida was claiming to want the same sacramental power as a priest," Ott said. "The level of dysfunction in the priesthood makes me question if it's right at all. . . . I'm skeptical of a woman who wants to get in. The whole business is about setting yourself above and apart; the priesthood as it stands now reifies the hierarchy. Maybe if enough women get into it, it will change. If women just go ahead and say we're in, it won't change." To her mind, such women lose all power because they are "dismissed."

Kretzu said she got three private e-mails in support of Raming's doing the liturgy. She suggested the women share their thoughts with the other Catholics but none did. The group agreed to have a meeting where they discussed the issue, but the controversy continued to swirl. A turning point came when Luis Ramos wrote Kretzu to say he had changed his mind; he now supported Raming's celebrating a Mass. It fell to the two of them to make the final decision. Kretzu says she was ready to chuck the whole idea of a liturgy, until she spoke to her mentor, black liberation theologian Dr. James Cone. Kretzu told him, "It's too much. I can't handle it. Everyone's going to hate me." But he wasn't having any of that. "So you sold out?" he said to Kretzu. His comment took her aback. She asked if he thought she should do it; he said yes. His input, plus the encouragement of several Protestant faculty members, convinced Kretzu to go ahead with the liturgy. They went with Ramos's suggestion. In deference to those Catholics who did not want Raming to do the liturgy, Raming would celebrate Mass outdoors at a nearby park, following her lecture in Union's Saint James Chapel. That's not what Kretzu would have done had it been up to her alone, but she appreciated the compromise.

Lighting Up the Night

Raming's lecture was scheduled to begin at 7:00 P.M. Until that very day, Kretzu had been knee-deep in planning the liturgy. "We had to get park permits . . . the lighting, the sound system, the altar set up out there . . . the choir, print up programs. It was so much work. I put more time into that than . . . in my own wedding," she tells me. At quarter to six, Kretzu was wielding a broom, sweeping the makeshift altar in the park, with barely enough time to shower. She had to race over to the chapel to deliver the welcome.

Moments before the lecture was to begin, Kretzu arrived. As she approached the podium, a tour organizer asked if she had an introduction written. For a moment, Kretzu panicked; she thought someone else was preparing it for her. She would have to wing it. Kretzu stepped up to the microphone, which wasn't working; she had to yell. She thanked the audience, some 175 people, for coming, and resorted to reading the flyer she had written on Raming's background. Then she began to ad lib, thanking Raming for coming, talking about Raming's life, what she had done, about the ordinations, about what they meant. Suddenly, she was overcome with emotion. "She has worked *for over forty years,*" said Kretzu, tearing up. "We just want to thank you *so much* for what you've done for our generation," she added, before breaking down. The audience rose to their feet and clapped. As for Raming, "she's so humble," recalls Kretzu, "that *she* stands up," saying, "Thank you, thank you."

Raming delivered her speech in her deliberate and plodding English, interrupted many times by applause. She linked the misogynistic beliefs of the Church fathers about women's innate inferiority, uncleanliness, and inability to image Jesus, which support the Church's refusal to ordain women, to the current representations of those beliefs. "For today, the continuing discrimination against women is dressed up and glossed over by Church officials," she said. "Now they speak of the 'genius of woman,' of her 'dignity,' of 'equal value but difference in kind.'" What it all comes down to, maintained Raming, is an argument for assigning women and men different roles and for ensuring that the roles assigned to women are the subordinate ones. She said: "It is time that this web of discrimination . . . is uncovered and torn asunder." Raming described the ordinations of the seven women on the Danube as a "prophetic sign of protest." She took issue with her excommunication, saying it is not excommunication by the Catholic community and certainly "this excommunication is not God's."

At the end of her presentation, a young man rose to speak. It was Luis Ramos. "I am very sorry and angered that you were excommunicated and made an exile, a stranger in the Church," he said. "I call out to Pope John Paul II to repent for allowing sexism and segregation to reign supreme in our Church. I call out to Cardinal Joseph Ratzinger to repent for excommunicating seven Roman Catholic women who were answering a holy calling. I repent for being silent in the midst of sexism in the Church."

After the lecture, Raming disappeared into a tiny ladies' room near the chapel. She entered in her street clothes but emerged fully vested in white chasuble and stole. I stood nearby, waiting to walk with her in the rain to Sakura Park on Riverside Drive, along the Hudson River, where she would celebrate the Eucharist. I was moved by the change not only in her clothes but in her demeanor. She wasn't "Ida" to me at that moment. I searched my mind for another way to address her. Father, of course, was inappropriate. Mother, by which women ordained in other denominations are addressed, felt awkward. Somewhat self-consciously, I greeted her as "Reverend Raming." "No," she said firmly. "I am Ida. You know me." Her words carried an unexpected resonance for me. "I am the good shepherd," Jesus said. "I know my own, and my own know me" (John 10:14).

Raming walked in silence amid a gaggle of chatting organizers to the park. They carried candles that managed to stay lit as we walked the two blocks in a light rain. When we arrived, Raming struggled to climb the steep stone stairs. Some seventy-five people, by my estimate, awaited her. The Mass was not polished. There were awkward moments and missed cues. Raming read from loose pieces of paper. But her audience was rapt. There were readings from the Bible—Acts of the Apostles, Wisdom, and John's Gospel; traditional prayers; and Communion. Raming consecrated the bread and wine before us, the makeshift altar illuminated gold against the black sky. A hush fell over the small crowd, then people walked one by one down an imaginary aisle in the grass to take the Communion from Raming's hands.

Speaking of the Mass afterward, Kretzu said: "The Episcopal women, when they were pushing for ordination, they did the same thing that Ida did. They went out and got ordained." She believes that the way the Episcopal Church moved from those irregular ordinations to the acceptance of women priests was that other Episcopal priests gave the women venues to celebrate Mass. They affirmed the women, refusing to "let the conversation die."

"Here's Ida," Kretzu continues. "She's worked all these years for this. She's struggled, written her thesis, submitted all these papers to the Vatican. Finally, she's like, I'm not getting any younger. I'm going to go out and do it." Raming is ordained, and then, says Kretzu, "she comes back to us, the community of women," asking for our support, believing that "that's how we're going to see justice in our Church that we love so much."

"Those women at Union who were saying 'no' to the liturgy were saying 'no' to women's ordination, period," insists Kretzu. "I don't care if they want to see women at some time ordained. This is your moment. It doesn't get more clear than this. Here is the opportunity. Are you going to do it or not? And they said 'no.' . . . I didn't have the guts to look at Ida do the lecture and say, 'Sorry, can't do the liturgy,' when she's gone out on that limb for us. She did it for her. But she did it for us."

Raming's next stop was Philadelphia. In a photo accompanying an Associated Press story, not one but seven Catholic women were on the altar during a Mass for hundreds of people gathered in the Friends Meeting House. Included were Raming and Iris Müller; Mary Ramerman; another newly ordained Spiritus Christi priest, Denise Donato; and the Reverend Judith Heffernan, one of hundreds of Catholic women in the United States ordained by their Catholic, small faith communities. While there is a very, very long way to go, it looked, in sociological parlance, like the beginning of critical mass—an essential element of social change.

Within the next year, two of the women ordained on the Danube, Christine Mayr-Lumetzberger and Gisela Forster, were ordained bishops, in Munich, Germany, by Catholic bishops whose identities were kept secret. The new women bishops, in turn, ordained six Catholic women deacons in another ceremony on the Danube, in 2004. More of these ordinations are planned for the United States in 2005, the thirtieth anniversary of the Women's Ordination Conference.

Catholic Women and Ordination: At the Crossroads

History is unlikely to judge kindly the Catholic hierachy's unrelenting condemnation of women who long to be priests in a Church that is literally crying for ministers, whose very future hangs in the balance. Yet, as the conflict at Union Theological Seminary illustrated, many Catholic women do not believe that women's ordination is *the* answer or even an answer to women's place in the Church or its current woes. Along the way, some of the most intelligent, accomplished, and passionate women reformers gave up the whole idea of fighting to get women admitted to the priesthood as it currently exists. Looking at a Church they see as hopelessly clerical and hierarchical, they want to transform that institution into what it once was, in the earliest days of Christianity—a discipleship of equals. They want what they have called from the birth of the women's ordination movement in America, "a renewed priestly ministry."

They remain, however, vague about what that new priestly ministry will look like. They completely reject the elevation of clerics over the laity. They want validation for all the diversified ministries that exist today. They see the priestly role as "no more elevated than the person who teaches children about Jesus, the person who collects food for the pantry, or the person who runs the parish council," said one reformer. They reject mandatory celibacy. If they envision ordination at all, it is an utterly democratic gesture, open to all—women and men, married and single, gay and straight, young and old—equipped with the knowledge and experience necessary to assume many different ministerial, as well as theological, roles.

Some who reject ordination as a major issue for Catholic women note that the ordination of women in other denominations has not solved the fundamental problem of imbedded sexism, of persistent opposition to women's full and equal participation in mainstream religions. Many ordained women wait but are never given a parish to pastor. Few head the largest and most prestigious parishes. Though the Reverend Joanna Adams was assigned to copastor a premier Presbyterian parish in Chicago in 2001, she stepped down three years later in the face of unending resistance to her authority. And many Catholic women reformers maintain that some women ordained in other denominations "are the most clerical people around," insisting on being called "Mother" all the time, living in their clerical collars, and setting themselves apart in much the same way that male clerics do. Catholic women in ministry have felt the sting of being turned away from gatherings of women ordained in other denominations because they have not been ordained themselves.

Some Catholic women who have given up waiting for ordination continue to turn to women's liturgy groups for spiritual sustenance. And women's liturgies remain central spiritual events at gatherings of progressive reform groups today. In the summer of 2003, at the twentieth anniversary celebration of the Women's Alliance for Theology, Ethics, and Ritual (WATER), there was a jubilant women's liturgy. After the meeting ended, a room full of women read psalms, prayed, sang, and joined hands in a spirited dance to the lyrics of "Sister Carry On." I loved seeing internationally renowned theologian Elizabeth Schüssler Fiorenza participate with such gusto in that circle, but I have to admit, I felt self-conscious, as if it were the 1980s again—and not in a good way. I never was much for those touchy-feely feminist encounters. I'm used to the reserve of the Catholicism I grew up with, the quiet worship, the depth of contemplation, the tradition of the Eucharist and the Mass that go back centuries. While I understand how women's liturgy groups can fill deep needs for some women, and I can see such liturgies supplementing the sacramental services of my Catholic faith, I cannot see those liturgies replacing them.

The vivacious and hardworking twenty-six-year-old director of the Women's Ordination Conference, Joy Barnes, feels the same way. She sees such rituals as "a little bit out there." She's looking for a different spiritual revolution. She tells me she's going to be married. She's thrilled that a Catholic woman minister active in WOC is going to celebrate the wedding along with her parish deacon. Right now, that is as close as Barnes can get to what she really wants, which is simply an ordained Catholic woman on the altar. "That's what we young people want," she tells me. "We want to change everything so that it can still be the same." I know exactly what she means.

Reformers make clear that to think that because Catholic women are not ordained, they are not achieving anything is a great misconception. I

agree. Clearly, women's leadership even without ordination has grown exponentially in the Church, and Catholic women ministers today are providing
a new and vivid model of leadership. However, if women are allowed to minister only within indefensible limits, only in ways that keep their true contributions hidden, and only in ways that underscore their second-class status,
then I fear the women will be doing more to uphold a dysfunctional institution than to midwife its birth into something better.

I love the idea that vocations are not just for priests anymore, that they
now are broadly defined and embrace the full range of Catholic ministries.
But I do not think we can stop there. I believe that the ordination of women
as deacons, priests, and bishops is essential if the Church is to recognize
women's spiritual authority and moral agency and if it is to insure women's
equal claim to ecclesial power and influence. Only as part of the Church's
governing structure can women—and this would be women from all sides
of the Church's political spectrum—help determine the shape of the future
Church, the participation of the laity, the selection of leaders—including
the next pope—and the interpretation of doctrines, especially those that
dictate the most private aspects of women's sexual, marital, and maternal
lives.

Validation by the Church of all the varied ministries in the world that
women occupy will not dismantle clericalism unless there is at first a level
playing field. Without that, women serve solely at the pleasure of clerics,
who can end those ministries on a whim, erasing once again women's central
roles in the Church. Without that, the Church will maintain its landed gen-
try of unmarried men with a stranglehold on the sacred whose task is to
oversee the minions—though I expect that will become ever more difficult
and desperate as the hierarchy's arguments for shutting women out ring
more and more hollow.

It is my feeling that once sacramentality is open to all, and all of the
Church's ministries are equally valued, work toward the truly servant priesthood called for by Christ can begin. That priesthood will no longer embody "power over" but "power to"—that is, the power to create a new model
of Church in the spirit of equality, mutual respect, open dialogue, and true
collaboration.

EPILOGUE: SAYING GOOD-BYE TO AUNT ANNE

In my last conversation with my beloved Aunt Anne, we sparred about feminist theology. An academic devoted to literature and the arts, as well as an artist and a musician herself, Aunt Anne had begun to read Elizabeth Johnson's *She Who Is*, at my suggestion. A deeply devoted Catholic, Aunt Anne called me in an agitated state. Why, she demanded, do all these women—like Johnson—spend so much time doing this research and writing? Plenty of Catholic women like things the way they are. And besides, nothing is going to change until there is another pope. Period.

I hadn't expected that, but indeed, throughout the two years it took me to write this book, I have thought over and over, why do these women stay? I asked that question of everyone. All of them talked in much the same terms about their love for the Church—the Eucharist and the Mass, the rituals and traditions, and the enduring commitment to justice and peace in the world—to which Sister Joan Chittister has devoted much of her energy these past two years. But beyond that, there were rich and varied responses.

Those who work on the inside persevere, believing that change depends on their staying. "Change isn't going to happen from me just quitting my job and becoming Mary Daly," said a chaplain. They persevere despite the risks. A young parish worker and Church activist knows that a day may come when she has to take a bold stand. "If I get thrown out for it," she says, "so be it."

Some stay out of sheer belligerence, unwilling to surrender the Church to its fundamentalist wing. Some stay specifically for other Catholic women. "There are lots of institutions women are voiceless in because they left," said one. Others are just plain obstinate. "I'm not going to be driven out of a Church that I love," said a theologian. "That would be a victory for the powers-that-be. I'm too tired, too stubborn to let that happen." Some stay to curb the Church's negative influence in the world on matters like reproductive health, while others stay to multiply the Church's protective voice for the poor, the powerless, and peace. Some stay because they believe the horrors of the clergy sex abuse crisis made them different, and they think that their being different will help to change the Church. "We're not going to be the same anymore," said a pastoral counselor. "We're not going to be so easily fooled. This will affect our daughters. They'll see us relate differently to the hierarchy, speaking out about things we never spoke out about before. We won't stay in our place anymore."

Some see the Church as a dysfunctional family that you "don't just abandon. You love them, no matter what. At the same time, you don't dream as much. You get real much faster." Some stay to preach change from their pulpits, as a lesbian pastoral assistant does for her largely gay congregation: "Like Jesus," she said one Sunday, "we, too, are at odds with some of the religious, political, and social teachings of our day. . . . And our challenge is the same as Jesus' challenge, to remain faithful to God, to what we know in our hearts to be true, no matter who or what institution demands differently."

Some are able to stay by separating their goals for the Church from their own spiritual lives. "I started out working for change in the Church, but that is not so much my goal anymore," said an older chaplain. She's looking for ways "to worship, to deepen my spirituality, and create opportunities for others to do likewise, whether it be with the Church or in spite of the Church." A theologian told me: "I'm a Roman Catholic because I believe in miracles." To her, the convener of the Second Vatican Council, Pope John XXIII, was a miracle, and she is waiting for another one. In the meantime, she moves on, "seeing how I can find the faith to do my work."

Some women, more than I expected, are hanging on by their fingernails. Highly trained women with a call to the priesthood are invited over and over to join the clerical ranks of other churches that accept women's gifts. Those who suffer the anguish and decide to leave are often given a send-off by Catholic women who have mentored them, who know the women must go but who watch with heavy hearts as they walk away. Some are more discouraged about the prospects of change than ever before, while others admit that if faced with the decision to choose a Church to join right now, it would not be the Catholic Church. Some are so aggrieved by the sex abuse scandals that they just don't know if they can stay. "The depth of my rage is only equal to the depth of my love for the Church," said a theology professor and activist.

And some change-makers have finally settled in, after years of personal struggle. "I feel like I spent so many years so angry," said a young chaplain. "Anger just takes so much time. . . . I wish I hadn't worried so much about where my road was going to lead me. I wish I had trusted more." On the subject of anger, the best advice I heard for change-makers came from sharp, ninety-year-old Janet Kalven. She was talking about the Grailville community that she has headed for more than sixty years, a community on a farm in Ohio that began as a meeting place for lay Catholic women, including the earliest feminist theologians, and has grown into a multipurpose spirituality and environmental education and retreat center. Those who live at Grailville and run the programs have worked hard, explains Kalven, to turn out women who are "successfully maladjusted, who don't settle for the status quo, who are determined to make changes, but without bitterness."

And staying means different things to different people. Mary Ramerman believes she has stayed. So does Ida Raming. So do many Catholic women active in women's liturgy groups and in small faith communities. What that speaks to is the contention of many reformers that there are many ways to be Catholic. In that regard, Ramerman's suggestion is well taken that what Catholicism needs is what Judaism has—room for everyone, in its Reform, Conservative, and Orthodox wings.

Aunt Anne never got to see these answers or read this book. Ever the professor, she said in our last conversation that she would finish Johnson's *She Who Is*, review it, and send me her thoughts in writing. I was amazed by the magnitude of that commitment and grateful. But then, she said she had to run.

Several months later, I got a call from her daughter, Margaret. Very suddenly, Aunt Anne, who had a chronic but not life-threatening condition, had come down with pneumonia. She and Margaret, a physician, were in Philadelphia at the time. Aunt Anne had an apartment there. They went to an ER, where Aunt Anne received medication. She was doing well enough to fly the next morning to her other home in Florida. Once there, she did better. But then, a few days later, she spiked a fever of 103. The trip to another ER turned into a hospitalization. The pneumonia raged. Her lungs filled with fluid. The antibiotics failed. On the fourth day in the hospital, my Aunt Anne died.

I could barely listen after that. But when the time came, my aunt was ready, Margaret said. They talked and laughed and told each other everything they needed to tell each other. They shared love and found comfort. Margaret and Uncle Frank were able to stay at Aunt Anne's bedside the whole time. They ate there. They slept there. At one point, looking at my sadly worn-out uncle, a nurse said: "You really should go home for a while." To which my dear aunt replied: "This man has been at my side for sixty-two years. He's not going to leave me now."

Anne and Frank met when they were twelve and thirteen respectively. They married young, had three children, years of joy, but also years of terrible heartache. Their son, Mark, an acclaimed concert violinist at the age of twelve, developed a form of muscle cancer that took his life at fourteen. Years later, another of Aunt Anne and Uncle Frank's children, Renee, a free spirit who lived in New York and traveled the world, walked outside into the night and vanished. Those losses left my aunt bereft, carrying a burden that never eased. But she did not lose her faith.

After Aunt Anne died, as Margaret was closing her eyes, there appeared in the doorway the hospital's Catholic chaplain. "She came in just at that moment," Margaret told me, marveling. They said the rosary together, "the whole rosary," which meant the world to her and my uncle. Aunt Anne had been anointed by a priest, but what the chaplain did for my cousin and my uncle struck me as a bridge back from death to life, a comforting extension of the hand of God. But it's interesting; even though I was writing this book at the time, when Margaret mentioned the Catholic chaplain, I immediately envisioned a priest.

On the day of the funeral, we filed up the marble stairs of Saint Lucy's Church in Scranton. This was the place where Aunt Anne's mother, my maternal grandmother, had worshipped, where she had became a friend to Mother Cabrini—later Saint Frances Cabrini—namesake of the elementary school next door. This is where we came for funerals but also for joyous occasions—confirmations, communions, weddings.

Approaching the brass doors before us was Aunt Anne in her casket, draped in an embroidered ivory satin cloth and carved with the Last Supper and two small *Pietas*. Aunt Anne was a lover of sculpture and a sculptor herself. In fact, she had fought tooth and nail for the cemetery where Mark is buried to mount a statue of Mary at his grave. The cemetery powers-that-be told her no more statues, but she persisted relentlessly until they finally gave in. Then she hired a famous sculptor to create it. Mark's statue is a tall, simple, sleek Mary with her arms outstretched to Mark. And now, to Aunt Anne.

My mother and I walked sadly down the aisle to our seats. We sat behind Uncle Frank and cousin Margaret, the remains of their little family. A young priest officiated. He was bright-eyed and energetic, and his sermon moved me deeply. He compared Aunt Anne's journey here to Jesus' journey in that they were teachers both. He called her an accomplished woman, a holy woman, a woman who lived her life the best way she knew how. And instead of dwelling with us in sadness at our loss, he took us to a place where we could watch her soar. He talked about the soul, about the earth and the stars and God's love and Aunt Anne's place in God's heart. There was Communion, which I took, and there were beautiful songs, selected by Margaret and Uncle Frank and sung by an extraordinary female cantor.

After the Mass, the priest sat down in a chair on the altar. The rest of us lowered ourselves onto the wooden pews. The other priest who had been officiating in his white alb walked to the side of the altar and picked up his viola.

Aunt Anne had been a public school music teacher for a time. This man as a young boy was one of her students. When he was in the seventh grade, Aunt Anne placed in his hands his first viola. The young man fell in love with the instrument and went on to become an accomplished musician as well as a priest. When he read in the local paper that Aunt Anne had died, he tracked down Uncle Frank and said he wanted to concelebrate the mass at Saint Lucy's, which pleased Uncle Frank and Margaret greatly. He had come to the wake, too, a tall lumbering man of humble bearing, his sadness apparent in the bow of his head and the look in eyes.

The Church went completely silent. This priest walked quietly down the steps from the altar. He stood among us, at the foot of Aunt Anne's casket. There, he played the "Ave Maria."

The woeful strains of his farewell floated over us. It was the ultimate act of homage, of gratitude, and of love. We wept, all of us, for Aunt Anne, for the human condition, and for the hope that God has given us in the persons of our ministers, our music, one another, and this Church.

I am Catholic still, I see. With all my hurt and all my anger, I am Catholic still. Because of the love. Because of the hope. Because of the community. And, oh. Because of the beauty.

NOTES

Introduction

ix. "We shall honor . . .": Elisabeth Schüssler Fiorenza, "The Intersection of Feminism and the Church," *Women Moving Church* (Center of Concern, Washington, D.C., 1982), p. 25.

xii. "On the Collaboration of Men and Women in the Church and in the World . . .": Letter to the Bishops of the Catholic Church, Congregation for the Doctrine of the Faith, approved by John Paul II, May 31, 2004; "liberation from biological determinism . . .": ibid., no. 2; "subordination because it causes . . .": ibid., nos. 2, 13.

xiii. "Religion and the Feminist Movement . . .": For voices from the conference, see Ann Braude, ed., *Transforming the Faiths of Our Fathers: Women Who Changed American Religion* (Palgrave Macmillan and the Harvard Divinity School Women's Studies in Religion Program, 2004).

1. The Revolt of the Erie Benedictines

1. "Dangerous Talk . . .": David Van Biema, "A Nun's Dangerous Talk," *Time*, August 20, 2001.

9. "We are . . . descendants . . .": From Sister Joan Chittister's speech "Heart of Flesh: A Feminist Spirituality for Women and Men" (see also her book by the same name, Wm. B. Erdmans Publishing Company, Grand Rapids, Michigan, 1998), presented at the meeting of SEDOS (Service of Documentation and Study in Mission) in Rome, Italy, December 5, 2000.

11. "When your most sublime ideas . . .": From Sister Joan Chittister's Closing Address "Leading the Way: To Go Where There Is No Road and Leave a Path," presented at the National Catholic Education Association meeting, in Milwaukee, Wisconsin, April 20, 2001.

14. "the 1,500-year-old rule of Saint Benedict . . .": Joan Chittister, *The Rule of Benedict: Insights for the Ages* (Crossroad, New York, 2001); "Benedictine authority and obedience . . .": Sister Christine Vladimiroff, "Background Information on

Deliberations with the Vatican," public statement posted on Erie Benedictines website, July 8, 2001.

16. "After much deliberation . . .": ibid.

16. "Sister Joan Chittister, who has lived . . .": ibid.

18. "Christian discipleship . . .": From Sister Joan Chittister's speech "Discipleship for a Priestly People in a Priestless Period," presented at the Women's Ordination Worldwide first international conference in Dublin, Ireland, June, 30, 2001; "Men who do not . . .": ibid.; "Only in the most backward . . .": ibid.

2. The Church and Reform: A Woman's Place

20. "Through cities and villages . . .": Luke 8:1–5.

21. "In return, women devoted . . .": From "A Call for National Dialogue on Women in Church Leadership" packet, a project of FutureChurch and Call to Action, see especially "Jesus and Women," "Women in the New Testament," "Women in the Hebrew Scriptures," "Women Ministers in the Early Church," and "Mary of Magdala" (FutureChurch, Cleveland, Ohio), *www.futurechurch.org;* "But the early Christian Church . . .": ibid.; "Clearly propelled . . .": Jane Schaberg, *The Resurrection of Mary Magdalene* (Continuum, New York/London, 2002), p. 82.

22. "Monthly defilement . . .": Ida Raming, *The Exclusion of Women from the Priesthood: Divine Law or Sex Discrimination,* translated by Norman R. Adams, (Scarecrow Press, Metuchen, New Jersey, 1976), p. 25; "the apostle Junia . . .": Barbara Kantrowitz and Anne Underwood, "The Bible's Lost Stories," *Newsweek,* December 8, 2003, p. 52; "Augustine believed . . .": Raming, *The Exclusion of Women from the Priesthood,* p. 36; "the Beguines . . .": Barbara Ballenger, "The Beguines: A Movement of Women and Spirit," in "Celebrating Women Witnesses Project" packet (FutureChurch, Cleveland, Ohio) and presentation by Edwina Gately at a Call to Action workshop, 2001 (see Chapter 11, "Catholic Women Raise the Dead," for more on Gately, the Beguines, and the workshop).

23. "Pray for others with unclean . . .": Uta Ranke-Heinemann, *Eunuchs for the Kingdom of Heaven: Women, Sexuality, and the Catholic Church,* translated by Peter Heinegg, (Doubleday, New York, 1990), p. 103; "love their wives as if they were sisters . . .": ibid., p. 105; "In the eleventh century . . .": ibid., p. 110; "Young husbands . . .": ibid., p. 107; "In 1139 . . .": Maureen Fiedler and Linda Rabben, editors, *Rome Has Spoken: A Guide to Forgotten Papal Statements, and How They Have Changed Through the Centuries* (Crossroad Publishing Company, NY, 1998), p. 130; "eight

subsequent popes . . .": "A History of Celibacy in the Catholic Church," original brochure developed by CORPUS Canada, revision by Call to Action and FutureChurch, December 1995; "Popes referred to priests' wives . . .": Ranke-Heinemann, *Eunuchs for the Kingdom of Heaven*, p. 111; "they were refused a church burial . . .": ibid., p. 112; "inspect the houses . . .": ibid., p. 115; "in Bamberg . . .": ibid., p. 116; "Up until the eighteenth century, abbesses . . .": Ruth A. Wallace, *They Call Her Pastor*, (State University of New York Press, 1992), p. 4.

24. "Of course, there has been progress . . .:" "Celebrating Women Witnesses" publications, including Catherine Meade on "Catherine of Siena," Martha Marie Campbell on "Teresa of Avila," Heidi Schlumpf on "Joan of Arc," and Robert McClory on "Mary Ward"; see also Joan Chittister, *A Passion for Life: Fragments of the Face of God* (Orbis Books, Maryknoll, NY, 2001), for profiles of Catherine of Siena, Teresa of Avila; "Catholic lay movement . . .": Carmel McEnroy, *Guests in Their Own House: The Women of Vatican II* (Crossroad Publishing, New York, 1996), pp. 26–27 regarding the first and second World Congresses for the Lay Apostolate; "no female face . . .": ibid., p. 45.

25. "with respect to the fundamental rights . . .": *Pastoral Constitution on the Church in the Modern World*, *(Gaudium et Spes)*, promulgated by Pope Paul VI, December 7, 1965, no. 29; "The possession of rights . . .": *Pacem in Terris*, "On establishing universal peace in truth, justice, charity, and liberty," encyclical of Pope John XXIII, April 11, 1963, no. 44; "Since women are becoming . . .": ibid., no. 41.

26. "*sensus fidelium* . . .": *Dogmatic Constitution on the Church (Lumen Gentium)*, promulgated by Pope Paul VI, November 21, 1964, no. 12. "Among the goals . . .": Summarized from "A Call for Reform in the Catholic Church: A Pastoral Letter from Catholics Concerned about Fundamental Renewal of our Church," first printed by Call to Action in the *New York Times* in 1990, with 4,505 signatures; "Women Doing Theology . . .": Janet Kalven, *Women Breaking Boundaries: A Grail Journey—1940–1995*, (State University of New York Press, 1999), pp. 209, 210, 225, and 227; "Mary Daly . . .": Mel Steel, "Mary, Mary, quite contrary," the *Guardian* (London), August 26, 1999; Lucy Hodges, Interview with Mary Daly, *The Times Higher Education Supplement*, June 7, 1996; Patricia Holt, "Arduous Journey for a Radical Theologian," *San Francisco Chronicle*, January 10, 1993; Michael Bronski, "Mary Daly vs. Boston College," Z magazine, October 1999.

27. "eleven female Episcopal deacons . . .": Carter Heyward, *A Priest Forever* (The Pilgrim Press, Cleveland, Ohio, 1976), from Preface, June 1998, page vii; "ratio of almost five to one . . .": *Women and the Catholic Priesthood: Proceedings of the*

Detroit Ordination Conference (Missionary Society of St. Paul the Apostle in the State of New York, 1976), p. 179; "Those women, who as teachers . . .": Elisabeth Schüssler Fiorenza, "Women Apostles: The Testament of Scripture," ibid., p. 100; see also: Schüssler Fiorenza, *Discipleship of Equals* (Crossroad: New York, 1998), pp. 211–232.

28. "After an unprecedented two years . . .": *Call to Action: 25 Years of Spirituality and Justice* (newsletter), July 2001, and Call to Action website, the 1976 U.S. Catholic Bishops' Call to Action Conference in Detroit Foundation documents, at *www.cta-usa.org/whobishconference/bishindex.html*; "Recommendations abounded . . .": ibid.; "The new way of doing the work . . .": *Call to Action: 25 Years of Spirituality and Justice.*

29. "a rebel worship service . . .": Marjorie Hyer, "Denied Robes, Women Hold Counter-Mass," the *Washington Post*, November 12, 1978.

30. "a renewed emphasis on cooperation . . .": Sue Costa, "Feminist Issue: Negative, Positive, and Future," in *Women Moving Church* conference proceedings (Center of Concern, Washington, D.C., 1982), p. 7; "A Catholic feminist spirituality . . .": Elisabeth Schüssler Fiorenza, "Getting Together in My Name . . . toward a Christian Feminist Spirituality," ibid., p. 11; "more than 5,000 women . . .": Diann L. Neu and Mary E. Hunt, *Women-Church Sourcebook* (WATERworks Press, Silver Spring, MD, 1993), pp 6–8; "Honoring Women's Blood Mysteries . . .": Diann L. Neu, *Women's Rites: Feminist Liturgies for Life's Journey* (The Pilgrim Press, Cleveland, Ohio, 2003), pp. 133–193; "A Ritual of Healing": Janet R. Walton, *Feminist Liturgy: A Matter of Justice* (The Liturgical Press, Collegeville, Minnesota, 2000), p. 61.

31. "A committee of six . . .": "Partners in the Mystery of Redemption: A Pastoral Response to Women's Concerns for Church and Society," *Origins*, Vol. 17, No. 45, April 21, 1988; "How can the hierarchical . . .": ibid., p. 766.

32. "An outcome unique . . .": Dolores Leckey, "Crossing the Bridge: Women in the Church, *Church*, Winter 2001; "natural family planning . . .": See "Contraception," in *Rome Has Spoken*, pp. 150 to 152, for history of Church's teaching on the rhythm method of birth control, beginning with its mention in *Casti Connubii*, promulgated by Pope Pius XI, 1930; "in 1976, Pope Paul VI . . .": *Report of the Pontifical Biblical Commission: Can Women Be Priests?* in "Rome's Main Documents on the Ordination of Women," *www.womenpriests.org/classic/appendix.htm*; "That same year, the Vatican . . .": *Inter Insigniores*, Declaration on the Question of Admission of Women to the Ministerial Priesthood, Sacred Congregation for the Doctrine of the Faith, October 15, 1976; "John Paul II would appoint . . .": Thomas J. Reese, S.J.,

"Notes on the October 21, 2003 Consistory," *America Press*, October 22, 2003 (*www.americamagazine.org/articles/consistory03.htm*).

33. "social, economic, cultural . . .": Pope John Paul II, "Letter to Women," issued prior to Fourth World Conference on Women in Beijing, June 29, 1995.

34. "if objective blame . . .": ibid.; "upholding the dignity . . .": ibid.; "Ordinary women . . .": ibid. "For in giving themselves . . .": ibid.; "mystery of women": ibid.; *Ordinatio Sacerdotalis*: "On Reserving Priestly Ordination to Men Alone," Apostolic Letter of Pope John Paul II, May 22, 1994; "still open to debate . . .": ibid.; "The next year, Cardinal Joseph Ratzinger . . .": Congregation for the Doctrine of the Faith, *Responsum ad Dubium, Concerning the Teaching Contained in Ordinatio Sacerdotalis*, October 28, 1995; "A few years later, the Vatican . . .": *Ad Tuendam Fidem (for the Defence of Faith), by which certain norms are inserted into the code of canon law and into the code of canons of the Eastern Churches*, promulgated by Pope John Paul II, May 28, 1998; *Commentary on Ad Tuendam Fidem by the Congregation for the Doctrine of the Faith*, Joseph Cardinal Ratzinger, June 29, 1998.

35. "for us and for our salvation . . .": The Nicene Creed, *Catechism of the Catholic Church*, (Image Book, Doubleday, New York, 1995) p. 56; "The issues are well known . . .": John L. Allen, Jr., "On the lectionary, eleven men made the deal," *National Catholic Reporter*, September 25, 1998.

36. "In the *Register* Sister Jeannine Gramick . . .": as quoted in an editorial "Catholics of Lincoln Deserve Better Than This," *National Catholic Reporter*, May 5, 2000; "Then eighty-four-year-old . . . :" ibid.; "advocate of witchcraft . . .": Rosemary Radford Ruether, "Thought control extends its reach in Lincoln," *National Catholic Reporter*, March 31, 2000; "pope-hating feminist . . .": Bob Reeves, "Catholics Urged to Avoid Meeting," *www.journalstar.com*, April 10, 2002; "Dominican Retreat House . . .": *www.lesfemmes-thetruth.org/2dominican.htm*; "She told the press that her artwork . . .": John L. Allen, Jr., "Bishop Shuts Down Women's Series," *National Catholic Reporter*, March 17, 2000; "If the water in the well . . .": Bishop Paul S. Loverde, "Bishop Cancels Speaker Series at Dominican Retreat House," March 3, 2000, *www.catholicherald.com*; "words of empowerment . . .": ibid.; "sound innocent enough . . .": ibid.

3. New Reformers, Old Reformers Face Off

40. "Only 20 percent of American Catholics . . .": William V. D'Antonio, James D. Davidson, Dean R. Hoge, and Katherine Meyer, *American Catholics: Gender, Gen-*

eration and Commitment (AltaMira Press, Walnut Creek, CA, 2001), pp. 78, 84; also appears in Angela Bonavoglia, "The Church's Tug of War," on women's role in Church reform, the *Nation*, August 19/26, 2002, adapted and expanded in this chapter.

41. "They referred to the *sensus* . . .": Voice of the Faithful documents, including "Our Rights and Responsibilities" and "Statement on Who We Are/Revised Initial Working Document," at *www.votf.org*.

43. "They drew up norms . . .": USCCB, Office of Child and Youth Protection, *Essential Norms for Diocesan/Eparchial Policies Dealing with Allegations of Sexual Abuse of Minors by Priests or Deacons*, approved December 8, 2002; "created a *Charter* . . .": "USCCB, Office of Child and Youth Protection, *Charter for the Protection of Children and Young People*, approved November 2002.

44. "clearly to promote the correct moral teaching . . .": "Statement by U.S. Cardinals," *New York Times*, April 25, 2002; "Cardinal Law forbade . . .": Letter from the Most Reverend Walter Edyvean, Moderator of the Curia, Archdiocese of Boston, on behalf of Cardinal Law, to priests, April 25, 2002; "VOTF called publicly . . .": Letter on Cardinal Law from Voice of the Faithful President, 12/9/02, *www.votf.org.*; "Cuenin . . . helped lead the drive . . .": Michael Paulson, "58 priests send a letter urging cardinal to resign," *Boston Globe*, December 10, 2002;

45. "Boston priest . . . leader in call for resignation," *Religion and Ethics Newsweekly*, Transcript: Episode no. 615, December 13, 2002; "theologian Reverend Richard P. McBrien . . .": "58 priests send letter urging cardinal to resign."

46. "Coming-of-age moment . . .": Troy, as quoted in James E. Muller and Charles Kenney, *Keep the Faith, Change the Church* (Rodale, 2004), p. 134.

49. "having her own 'agenda' . . .": Letter from Most Reverend Thomas J. O'Brien, Bishop of Phoenix, to Sandra M. Simonson, May 3, 2002; "Your letter indicates . . .": ibid.; "wrote another letter to O'Brien . . .": Letter to Bishop Thomas O'Brien from Sandy Simonson, on behalf of VOTF Arizona Leadership Team, June 8, 2003 (posted, *www.votf.org*).

50. "During Easter Week in 2004, Archbishop Sean P. O'Malley . . .": Michael Paulson, "Prelate disallows women in ritual—washing of the feet is limited to males," *Boston Globe*, April 10, 2004.

51. "Adding insult to injury . . .": Archbishop Sean P. O'Malley, *Chrism Mass Homily*, April 6, 2004, and Eileen McNamara, "Linking Evil to Feminism," *Boston Globe*, April 11, 2004.

56. "The truth is that . . .": Joan Chittister, "Voice of the Faithful goes after biggest issue of all: authority," *National Catholic Reporter*, January 31, 2003.

4. Sex, Priests, and Girlhoods Lost

60. "A Latina who grew up . . .": For full story, see Angela Bonavoglia, "The Sacred Secret," *Ms.*, March/April 1992, pp. 40–45.

62. "Cruces was discovered back in New York . . .": Alex Tresniowski, Rebecca Paley, John Hannah, Ron Arias, Frank Swertlow, Melissa Schorr, and Nelly Sindayen, "Unholy Fathers," *People*, June 3, 2002, p. 62; "Bishop Daily resigned . . .": Daniel J. Wakin, "Brooklyn Bishop Ending Tenure amid Storm over Scandal," *New York Times*, August 2, 2003; Tom Fox, "Daily's legacy tarnished by clergy sex abuse record," *National Catholic Reporter*, August 4, 2003; and Office of the Attorney General Commonwealth of Massachusetts, "The Sexual Abuse of Children in the Roman Catholic Archdiocese of Boston, p. 32–33; "She launched a website . . .": *http://faculty.uml.edu/sgallagher/spotlight.htm*.

63. "Of the 99 victims . . .": Sacha Pfeiffer, "Women Face Stigma of Clergy Abuse," *Boston Globe*, December 27, 2002; "Kelley admitted to . . .": *Worcester Telegram & Gazette*, May 11, 2002; "Carl Warnet preyed on . . .": See *60 Minutes II*, "The Church on Trial: Part 2," June 12, 2002, for more; "It revealed that more than 10,000 . . .": John Jay College of Criminal Jus-tice Survey, *The Nature and Scope of the Problem of Sexual Abuse of Minors by Catholic Priests and Deacons in the United States*, USCCB, February 27, 2004 (*www.usccb.org/nrb/johnjaystudy*), pp. ix, viii, based on data gathered from 195 dioceses and eparchies (the 17 Eastern rite Catholic Churches in the United States, e.g., Byzantine, Ukranian), and 140 religious communities, representing 97 percent of all diocesan priests in the United States and 80 percent of all religious order priests (e.g., Jesuits, Franciscans); "Nineteen percent of those victims were girls . . .": ibid., p. xi; "gender of victims was unknown . . .": ibid., p. 41; "26 percent either abused . . .": ibid.; "More than three quarters of the victims . . .": National Review Board for the Protection of Children and Young People, "A Report on the Crisis in the Catholic Church in the United States," February 27, 2004, p. 26 (*www.usccb.org/nrb/nrbstudy/nrbreport*); "The largest number of girls . . .": John Jay survey, p. 42.

64. "Based on that underreporting . . .": Mark Mueller and Jeff Diamant, "Critics Say Church Tally of Abuse Is Incomplete," *Religious News Service*, March 2, 2004; "Finally, the vast majority of cases . . .": John Jay survey, p. 17.; "plaintiffs' attorneys

report . . .": Thomas Farragher and Matt Carroll, "Church Board Dismissed Accusations by Females," *Boston Globe*, February 7, 2003.

65. "Nancy Sloan was . . .": See also Angela Bonavoglia, "New Battleground for Survivors of Priest Child Sex Abuse," *Ms.*, December 2002/January 2003, pp. 39–40.

66. "The developmental stage . . .": "Suffolk County Supreme Court Special Grand Jury Report, January 17, 2003, p. 98; "In a banner story . . .": Melinda Henneberger, "Pope Offers Apology to Victims of Sex Abuse by Priests," *New York Times*, April 24, 2002.

67. "He's not a predator . . .": Alison Leigh Cowan, "New Clergy Abuse Settlement is Announced in Bridgeport," *New York Times*, October 17, 2003; "In announcing the diocese's decision . . .": Press release by Dr. Joseph McAleer, Diocese of Bridgeport, November 22, 2002; "He is entitled . . .": ibid.

68. "Finally, in his great sympathy for an inebriated priest . . .": John Jay survey, executive summary p. x; "It's almost a free pass . . .": Farragher, "Church Board Dismissed Accusations by Females."

69. "Sweeney was 'a practicing Catholic . . .' ": Kevin Cullen, "Crisis in the Church; Seven People Who Made a Difference," *Boston Globe*, December 15, 2002; "on a tour of the corridors . . .": Mary Gail Frawley-O'Dea, "The Experience of the Victim of Sexual Abuse: A Reflection," speech presented to the USCCB, June 14, 2002, p. 3; "incest . . . reflect on your role . . .": ibid., p. 1; "I have a lot of empathy . . .": Michael Paulson, "Abuse specialists challenge church defense tactics," *Boston Globe*, January 22, 2003.

70. "Frawley-O'Dea joined a volunteer group . . .": Ralph Ranalli, "Group Offers to Review Abuse Cases," *Boston Globe*, October 9, 2003; Lisa Wesel, "Panel reviews sexual abuse cases," website of the Massachusetts Psychologist, August/September 2004 (*http//info@vrcboston.org*.); "to the consternation of some conservative . . .": Alan Cooperman, "Some Bishops Resisting Sex Abuse Survey," *Washington Post*, June 10, 2003.

71. "90 percent of the dioceses audited . . .": Office of Child and Youth Protection, *Report on the Implementation of the Charter for the Protection of Children and Young People*, USCCB, December, 2003, p. 9 (*www.usccb.org/ocyp/audit2003/report.htm*); "But McChesney built strong observations . . .": ibid., Chapter 4, Recommendations; "La Cosa Nostra . . .": Larry B. Stammer, "Mahony Resisted Abuse Inquiry, Panelist Says," *Los Angeles Times*, June 12, 2003; "troubling lack of any expres-

sion . . .": National Review Board Report, p. 94; "don't rock the boat . . .": ibid.,
p. 128; "outspoken priests rarely were selected . . .": ibid.

72. "the type of priests who are chosen as bishops . . .": ibid.; "The exercise of
authority . . .": ibid., p. 126; "In 2004, the board found out that behind their
backs . . .": Letter from Henry Mansell, Archbishop of Hartford, on behalf of Hartford bishops, to Wilton D. Gregory, USCCB president, February 12, 2004; Letter
to Wilton D. Gregory from 23 bishops from New Jersey and Pennsylvania; "In re-
sponse, Burke sent a blistering . . .": Letter from Honorable Anne Burke, Illinois
Court of Appeals, Justice and Interim Chair of the National Review Board, to
Wilton D. Gregory, March 30, 2004; "Chagrined and threatened, some bishops . . .":
Letter from Denver Archbishop Charles J. Chaput and Denver Auxiliary Bishop
Jose H. Gomez to Anne Burke, April 2, 2004; Letter from Charles V. Grahmann,
Bishop of Dallas to Anne Burke, April 6, 2004; and Letter from David L. Ricken,
Bishop of Cheyenne to Anne Burke, April 16, 2004.

76. "clearly a troubled . . .": Press statement of Reverend James F. Cryan, OSFS,
January 28, 2003; "Where do we place . . .": "Shame, Sin, and Secrets," *Toledo Blade*,
December 1, 2002.

77. "While more than half the states . . .": Bonavoglia, "New Battleground for
Survivors;" "Within that twelve-month window . . .": Jean Guccione and William
Lobdell, "California Law Spurred Flood of Sex Abuse Suits," Los Angeles Times,
January 1, 2004.

78. "tactics SNAP discovered . . .": Heidi Singer, "Church Hot-Line Furor,"
New York Post online edition, August 30, 2004; "And SNAP is ever on the lookout . . .": "Some Bishops Split Hairs to Protect & Keep Abusers in Active Ministry," SNAP Press Statements/Fact Sheet, July 29, 2003; "Suspected Abusers
Remaining in Ministry," A SNAP Fact Sheet, November 11, 2003; "priest in Santa
Ana, California . . .": "FBI investigates child porn allegations against Santa Ana
priest," Associated Press, web post July 26, 2003; "priest . . . living in Chicago's Car-
dinal Francis George's residence . . .": "Priest's case not covered by church abuse policy," *Chicago Sun-Times*, March 2, 2003; "joint meeting of representatives . . .":
Pre-Assembly Press Release, "Joint Conference of Major Superiors of Men–
Leadership Conference of Women Religious, July 27, 2004; "environment conduc-
tive to listening . . .": Leadership Conference of Women Religious press release,
Response of LCWR President Sister Constance Phelps, SCL to SNAP Request,
August 13, 2004.

80. "observers and participants alike . . .": "Cardinal leads prayer service with survivors," *The Catholic Review*, March 9, 2004; "no place in the priesthood . . .": Pope John Paul II, Vatican Information Service, April 23, 2002.

82. "Even the National Lay Review Board recommended . . .": National Review Board Report, p. 21; "Robert S. Bennett": Appearance on *Meet the Press*, February 29, 2004; "The Vatican supports easing off . . .": John Thavis, "Vatican: Church must work with scientific experts to prevent abuse," *Catholic News Service*, February 18, 2004.

83. "even the president of the U.S. Bishops' Conference . . .": Patrick J. Powers, "Former altar boy sues priest removed in 1995," *Belleville News-Democrat*, September 26, 2003; William Lamb, "Privacy of medical records is issue in case," *St. Louis Post-Dispatch*, May 5, 2004; SNAP press statement, "SNAP Comments Regarding Contempt Ruling Against Diocese," April 7, 2004; *Gina Trimble Parks et al., Appellees, v. Raymond Kownacki et al.*, Opinion filed August 10, 2000, FindLaw; "In 2003, Rockford, Illinois, Bishop . . .": Dan Rozek, "Diocese says guilty plea should end records case," *Chicago Sun-Times*, May 15, 2004; "Ex-priest pleads guilty to sexual abuse," WTVO Channel 17 (Rockford, Illinois), May 14, 2004.

84. "After a yearlong investigation of more than two hundred . . .": "Runaway priests hiding in plain sight," Special Report/Series, *Dallas Morning News*, June 21, 2004; "Overwhelming evidence . . .": Thomas F. Reilly, "The Sexual Abuse of Children in the Roman Catholic Archdiocese of Boston," A Report by the Attorney General, Office of the Attorney General, Commonwealth of Massachusetts, July 23, 2003; "four most important basilicas . . .": Al Baker, "Cardinal Law Given Post at Vatican," *New York Times*, May 28, 2004; "Even today, some bishops and priests fail . . .": National Review Board Report., p. 109.

85. "Many Church leaders . . .": ibid., pp. 98–99.

5. Women, Priests, and the Myth of Celibacy

86. "At a press conference in Manhattan . . .": A Call to Accountability, UN Church Center, July 14, 2001, www.calltoaccountability.org; "story broken by the National Catholic Reporter . . .": John L. Allen, Jr., and Pamela Schaeffer, "Reports of Abuse: AIDS exacerbates sexual exploitation of nuns, reports allege," National Catholic Reporter, March 16, 2001; "The reports documented . . .": "Personal Memo from Sr. Maura O'Donohue MMM, [re] Meeting at SCR, Rome, 18 February 1995, memo dated 21 February 1995; "Memo from Sr. Maura O'Donohue MMM: Urgent Concerns for the Church in the Context of HIV/AIDS," February

1994; and memo from Marie McDonald, MSOLA, "The Problem of the Sexual Abuse of African Religious in Africa and Rome," November 20, 1998, all published on the National Catholic Reporter website.

87. "Were forced into becoming a second . . .": O'Donohue, February 1994 memo, p. 5; "challenging their pastors . . .": ibid., p. 5; "the presbytery in a particular parish . . .": ibid., p 6.

88. "The Cannibal's Wife . . .": Yvonne Maes, with Bonita Slunder (Herodias, NY/London, 1999).

89. "Please join me . . .": Letter from Archbishop Renato R. Martino, Apostolic Nuncio, Permanent Observer of the Holy See to the United Nations, to Dr. Mary E. Hunt, Women's Alliance for Theology, Ethics, and Ritual, for the Call to Accountability Coalition, July 11, 2001; "The problem is known . . .": NCR staff, *National Catholic Reporter*, "Vatican acknowledges abuse of women religious by priests," (contains NCR translation of Vatican statement by papal spokesperson Dr. Joaquin Navarro-Valls), March 20, 2001; "An estimated half of all priests . . .": A. W. Richard Sipe, *Celibacy in Crisis: A Secret World Revisited* (Brunner-Routledge, New York and Hove, 2003), pp. 50–52; "numerous witnesses . . .": National Review Board Report, p. 88–89; "According to a recent *Los Angeles Times* poll . . .": Survey of Roman Catholic Priests in the U.S. and Puerto Rico, *Los Angeles Times* poll, June 27 to October 11, 2002, p. 12.

90. "Celibate law is extolled . . .": Richard Sipe, "Beyond Crisis," presentation at Sexuality and the Modern Priesthood workshop, Voice of the Faithful Regional Conference ("Being Catholic in the twenty-first Century"), Fordham University, Bronx, NY, October 25, 2003; "In 2003, plaintiffs' attorney . . .": Michael Rezendes, "Church to Disclose Records of Clergy Accused by Adults," *Boston Globe*, January 29, 2003; attorney Roderick Eric MacLeish, Jr., "Plaintiffs' Memorandum of Law in Support of Motion in Limine to Admit Evidence of Practices and Policies of the Roman Catholic Archbishop of Boston, Concerning Sexually Abusive Priests Other than Paul R. Shanley," submitted to Commonwealth of Massachusetts Superior Court Department, in *Gregory Ford, et. al., Plaintiffs v. Bernard Cardinal Law, et. al.*, July 21, 2003; "Rita J. Perry . . .": "James D. Foley" in "Plaintiffs' Memorandum," pp. 65–73; Stephen Kurkjian and Walter V. Robinson, "A 'Classic Misuse of Power' ", *Boston Globe*, December 29, 2002; Stephen Kurkjian, "Priest Details Longtime Affair Tells Four Children of the Woman Who Died That He's Sorry," *Boston Globe*, January 31, 2003; "DNA tests show priest is father of 2 children," Associated Press, October 29, 2003; and staff and wire reports, "Church settles wrongful death suit," Associated Press, January 30, 2004.

91. "A large amount of hair . . .": "Thomas P. Forry" in "Plaintiffs Memorandum," pp. 73–82; "Auxiliary New York City bishop . . .": "Auxiliary New York bishop resigned: Statement by McCarthy," *CNN.com*, June 12, 2002; "at least fourteen had been accused . . .": Brooks Egerton, "Policy Doesn't Deal with Priests Who Abuse Adults," *Dallas Morning News*, November 11, 2002; "Worldwide, 21 Roman Catholic bishops have resigned amid church scandals since 1990," Associated Press, June 18, 2003.

94. "no matter what the level . . . ," Peter Rutter, M.D., *Sex in the Forbidden Zone: When Men in Power—Therapists, Doctors, Clergy, Teachers, and Others—Betray Women's Trust* (Jeremy P. Tarcher, Inc., Los Angeles, 1989), p. 21; "Twenty-three states have laws . . .": Interview with Gary Schoener; see also Advocate Web: Helping Overcome Professional Exploitation, for civil and criminal codes and federal law, *www.advocateweb.org*.

95. "Psychologist Richard Sipe, too, has observed . . .": Sipe, *Celibacy in Crisis*, p. 125.

96. "An estimated twenty thousand . . .": "Celibacy Research," www.rentapriest.com/CITI.htm; "30 percent of priests involved with women . . .": Sipe, *Celibacy in Crisis*, p. 51; "Priests who had entered into civil marriages . . .": ibid., p. 117.

103. "Serious concerns . . .": Letter from the Los Angeles Province and Lay Associates of the Sisters of St. Joseph of Carondelet to Cardinal Roger Mahoney, August 3, 2003; "collaborative model of leadership . . .": Letter from Cardinal Roger Mahoney to the Sisters of St. Joseph of Carondelet and Lay Associates, August 14, 2003; "the most meaningful ministry . . .": "A Message from Sr. Judy Molosky, "We Are St. Brendan Church" newsletter, September 7, 2003; "Kathleen McChesney . . . argued for consistent . . .": *Report on the Implementation of the Charter for the Protection of Children*, p. 39, no. 1.5; "Bishops and other church leaders . . .": National Review Board Report, p. 88–89; "30 percent to more than 50 percent . . .": Sipe, *Celibacy in Crisis*, p. 51; Donald B. Cozzens, *The Changing Face of the Priesthood* (The Liturgical Press, Collegeville, MN, 2000), p. 99.

104. "Gay subculture . . .": National Review Board Report, p. 81; "Within the Church's distorted sexual teaching . . .": Sydney Callahan, "Stunted Catechesis: The Church's teaching on sexuality has a role in its current crisis," Boston *C21 Resources*, Boston College, Fall 2003, pp. 4–5 (reprinted from *National Catholic Reporter*, March 21, 2003); "bishops, provincials, and seminary rectors . . .": National Review Board Report, p. 81; "bishops chastised Kathleen McChesney . . .": Letter

from Henry J. Mansell, Archbishop of Hartford, on behalf of the bishops of Hartford, to Wilton D. Gregory, February 12, 2004; "to married as well as celibate men . . ." Letter from 163 priests in Milwaukee archdiocese to Wilton D. Gregory, August 18, 2003, as published by the *Milwaukee Journal Sentinel, www.jsonline.com*; "the ever growing appreciation of marriage . . .": ibid; "Mandatory celibacy is built . . .": Anthony Padavano, speaking at Catholics for a Free Choice press conference, "First Campaign Calling for the UN to Intervene in Clergy Sexual Abuse of Children," during UN Special Session on Children, New York, NY, May 8, 2002.

105. "The Church's Eastern rite priests . . .": *Catechism of the Catholic Church*, no. 1580, p. 440 (Doubleday, 1994); "Church began to allow in 1980 . . .": Document Outlining the Pastoral Provision issued by the Sacred Congregation for the Doctrine of the Faith on July 22, 1980, *www.marriedpriests.org/Provisions.html*; "On the other hand, many new seminarians . . .": Jerry Filteau, "Seminarians Show Support for Celibacy," *St. Louis Review Online*, August 27, 2004.

106. "The abuse of vulnerable . . .": Susan Archibald, "The Linkup, Inc.—Survivors of Clergy Abuse: Statement on the Compliance Audit of Roman Catholic Dioceses," January 6, 2003, Web posting.

107. "The magisterial teaching is that every . . .": Sipe, "Beyond Crisis."

6. The "A" Word, the "C" Word, and the Fight for Women's Moral Authority

110. "141 recipient countries . . . $59 million in funding . . .": For information on the United Nations Population Fund, go to *www.unfpa.org*; "for taking this action . . .": "Bishop Gregory praises decision to withhold U.N. population funds," *Catholic News Service*, July 24, 2002; "She gathered together a delegation . . .": Report of an Interfaith Delegation to China, *United Nations Population Fund in China: A Catalyst for Change* (CFFC, 2003), pp. 16–19 for counties visited, 30–34 for itinerary; "We believe that the work of UNFPA . . .": U.S. Religious Leaders and Ethicists Urge Senators to Vote for Full US Funding to UNFPA," CFFC Press Release, October 1, 2003.

111. "Unwritten law embedded . . .": Christine E. Gudorf, *Body, Sex, and Pleasure: Reconstructing Christian Sexual Ethics* (The Pilgrim Press, Cleveland, Ohio, 1994), p. 62.

112. "Most notorious anti-Catholic . . .": "60 Minutes Draws on Bigot to Slam Church," *Catalyst* (online), Catholic League, January-February, 2001; "an arm of the abortion lobby . . .": "National Conference of Catholic Bishops/United States

Catholic Conference President Issues Statement on Catholic for a Free Choice,"
May 11, 2000; "policies held by the Holy See . . .": Vatican's Pontifical Council for
the Family, "Catholics for a Free Choice" chapter, *Lexicon on Ambiguous and Colloquial
Terms About Family Life and Ethical Questions*, March 31, 2003.

113. *"The Choices We Made . . .":* Angela Bonavoglia, *The Choices We Made: 25
Women and Men Speak Out About Abortion* (Random House, 1991; Four Walls,
Eight Windows, 2001). An oral history featuring my interviews with mostly fa-
mous women (including Catholics) and two men about their personal experiences
with abortion, from the 1920s through the 1980s, with a foreword by Gloria
Steinem.

115. "CFFC in 1984" . . . : "A Diversity of Opinions Regarding Abortion Exists Among Committed Catholics," *New York Times* ad, October 7, 1984.

116. "Ninety-seven Catholics . . .": Mary E. Hunt and Frances Kissling, "The
New York Times Ad: A Case in Religious Feminism," *Conscience*, Spring/Summer
1993; "After being raked over the coals . . .": Barbara Ferraro and Patricia Hussey
with Jane O'Reilly, *No Turning Back: Two Nuns' Battle with the Vatican over Women's
Right to Choose* (Poseidon Press, 1990).

117. "Catholic social workers in Los Angeles . . .": Marjorie Hyer, "Liberal
Catholics Split: Ad Protests Reprisals for Abortion Dissent," *Washington Post*,
March 3, 1986; "Mary Ann Sorrentino . . .": ibid.; "basic right to life from conception . . .": Congregation for the Doctrine of the Faith, *Doctrinal Note on Some Ques-
tions Regarding the Participation of Catholics in Political Life*, approved by John Paul II,
November 21, 2002, published January 2003; "not fit to receive communion . . .":
Arinze as quoted in Laurie Goodstein, "Vatican Cardinal Signals Backing for Sanctions on Kerry," *New York Times*, April 24, 2004; "Cardinal Joseph Ratzinger . . . essentially agreed . . .": "Worthiness to Receive Holy Communion—General
Principles," *L'espresso*, June 2004.

118. "Bishops, task force developed a public statement . . .": USCCB,
"Catholics in Political Life," July 1, 2004.

119. "revive latent anti-Catholic prejudice . . .": Letter from 48 Democratic
Congress members to Cardinal Theodore McCarrick, Chairman of the Task Force
on Catholic Bishops and Catholic Politicians," May 10, 2004; "I'm not going to have
a fight . . .": Cardinal Theodore E. McCarrick as quoted in Jodi Evans, "The Politics
of Communion," *Conscience*, Summer/Autumn 2004, p. 15 (comment originally
made to Catholic Press Association on May 27, as per *Catholic News Service*); "poll of

likely voters release in 2004 . . .": *The View from Mainstream America: The Catholic Voter in Summer 2004*, conducted for Catholics for a Free Choice by Belden, Russonello & Stewart, July 2004; "a *Time* magazine poll": "U.S. Catholics Separate Politics and Faith," *Time*, June 21, 2004.

120. "Canon law does say that . . .": *Catholics and Abortion: Notes on Canon Law No. 1* (Catholics for a Free Choice, Washington, D.C., 2003); "Catholic women in 2000 accounted for 27 percent . . .": *Perspectives on Sexual and Reproductive Health*, Alan Guttmacher Institute, Volume 34, Number 5, September/October 2002 (*www.guttmacher.org/tables/3422602t.html*); "61 percent agreed that abortion . . .": *The View from Mainstream America*; "Hates being fat . . .": USCCB advertisement, "President Clinton wants the exceptions for legal partial birth abortion slightly broadened," *Washington Post*, March 25, 1996.

121. "96 percent . . .": "Catholics and Contraception: The Facts Tell the Story," Catholics for Contraception, based on data from the 1995 National Survey of Family Growth, National Center for Health Statistics, National Institutes of Health (*www.catholicsforchoice.org/contraception/catholics.htm*).

122. "many countries around the world . . .": *A World View: Catholic Attitudes on Sexual Behavior and Reproductive Health* (Catholics for a Free Choice, 2004); "'unitive' as well as the 'procreative' . . .": "Contraception" in Fiedler and Rabben, *Rome Has Spoken*, p. 156; see also *Humanae Vitae*; "merger mania. . . .": Cynthia Gibson, "Catholic Hospitals and Health Care Reform—The Big Business of Healing," *Conscience*, Summer 1994.

123. "The enemy . . . flesh of their own flesh . . ." Frances Kissling, "A Callous and Coercive Policy," Religion News Service, May 24, 1999.

124. "As much as 68 percent more . . .": Women's Research and Education Institute, *Women's Health Insurance Costs and Experiences*, 1994; "By late 2004, some 21 states . . .": "Insurance Coverage of Contraceptives," *State Policies in Brief*, The Alan Guttmacher Institute, October 1, 2004 (*www.guttmacher.org*); "standard I helped to develop . . .": Angela Bonavoglia, *Co-opting Conscience: The Dangerous Evolution of Conscience Clauses in American Health Policy* (ProChoice Resource Center, January 1999).

125. "Why should an entity . . .": as quoted by Laura Flanders, "Giving the Vatican the Boot," *Ms.*, October/November 1999.

126. "The human rights of women include . . .": *Beijing Declaration and Platform for Action*, Women and Health, no. 96; "When one reviews the history . . .":

Frances Kissling, "From Cairo to Beijing and Beyond," *Conscience*, Winter 1995/96, p. 20.

128. "In sub-Saharan Africa . . .": *UNAIDS 2004 Report on the Global AIDS Epidemic* (Joint United Nations Programme on HIV/AIDS, June 2004); "AIDS increasingly has a woman's face . . .": Kofi A. Annan, "In Africa, AIDS Has a Woman's Face," op-ed, *New York Times*, December 29, 2002; "Women make up 50 percent": *UNAIDS 2004 Report;* "women represent 30 percent": *HIV Prevention Strategic Plan Through 2005,* Centers for Disease Control, January 2001; "Women of color": ibid.; "anguish over the enormity of the suffering . . .": Statement by Bishop Kevin Dowling of Rustenburg, issued before the bishops' meeting at the end of July 2001, *www.sundaytimes.co.za*, July 9, 2001.

129. "widespread and indiscriminate promotion . . .": "A Message of Hope," from the Catholic Bishops of the People of God in South Africa, Botswana and Swaziland, Archdiocese of Cape Town, *catholic-ct.org.za*, July 30, 2001; "spermatozoon can easily pass . . .": Steve Bradshaw, Panorama reporter, "Vatican—condoms don't stop HIV," *BBC News*, October 10, 2003 *(http://news.bbc.co.uk/go/pr/fr/-/1/hi/programmes/panorama/3180236.stm);* "When priests preach against using contraception . . .": US-AIDS director Peter Plot as quoted in "Church's stand against contraception costs lives," *Agence France Presse,* June 29, 2001 (see *www. Condoms4Life.org,* "Pros & Cons: Voices on the Church's Condoms Policy); "These incorrect statements . . .": Bradshaw; "essentially impermeable . . .": "Workshop Summary: Scientific Evidence on Condom Effectiveness for STD Prevention," National Institute of Allergy and Infectious Diseases, National Institutes of Health, Dept. of Health and Human Services, July 20, 2001, p. 10; "reduces risk of HIV infection by at least 90%": "Making Condoms Work in HIV Prevention," Joint United Nations Programme on HIV/AIDS (UNAIDS), June 2004, p. 16.

130. "not an 'intrinsic evil' . . .": Bishop Kevin Dowling, "Let's not condemn condoms in the fight against AIDS," *U.S. Catholic,* November 2003, p. 21; "the beautiful act of love . . .": "Continuing the Conversation—A response to two recent statements—that of Bishop Kevin Dowling after the UNAIDS conference (9 July 2001) and the South Africa Catholic Bishops Conference "Message of Hope" (30 July 2001) on the ethics of using condoms to prevent the spread of AIDS," *www.sundaytimes.co.za,* August 16, 2001.

131. "I spent twenty years . . .": *Catalyst,* Catholic League.

133. "The pro-choice movement will be . . .": Frances Kissling, "Is There Life

After Roe? How to Think About the Fetus," *Conscience*, Winter 2004/5, pp. 10–18.

7. Strife in the Ranks: The Abortion Challenge to Church Reform

135. "We find no judgment on persons . . .": Christine E. Gudorf, "To Make a Seamless Garment, Use a Single Piece of Cloth," in *Abortion and Catholicism: The American Debate*, edited by Patricia Beattie Jung and Thomas A. Shannon (Crossroad, New York, 1988), p. 282; "to provoke reflection . . .": ibid.; "Why is persuasion? . . .": ibid.; "Women with the medical option . . .": ibid., p. 284; "I am not proabortion . . .": ibid., p. 279.

138. "68,000 . . . die from unsafe abortions . . .": World Health Organization, "Unsafe Abortions," *www.who.int/reproductive-health/unsafe_abortion/index.html*.

139. "You can believe . . .": Michele Curay-Cramer as quoted in Randall Chase, "Catholic Teacher Sues Bishop, Diocese," Associated Press wire story, November 14, 2003.

142. "Catholic social teaching is . . .": See Shaping a New World: A Challenge for the twenty-first Century (NETWORK Education Program, Sixth Edition, 1998) for useful summary.

145. "liberation from biological determinism . . .": "Letter to the Bishops of the Catholic Church on the Collaboration of Men and Women"; "how these two dimensions . . .": John Paul II, *Mulieris Dignitatem, On the Dignity and Vocation of Women, Origins*, Vol. 18, No. 17, October 6, 1988, p. 273.

8. Redeeming the Church's Sexual Outlaws

148. "Those who would move . . .": Congregation for the Doctrine of the Faith, *Considerations Regarding Proposals to Give Legal Recognition to Unions Between Homosexual Persons*, June 3, 2003, no. 5; "allowing children to be adopted . . .": ibid., no. 7; "if gays were allowed to wed . . .": Brooklyn Bishop Nicholas DiMarzio, in appearance on Fred Dickler's radio talk show, WROW-AM, Albany, New York, March, 2004, as quoted in Thomas J. Lueck, "Bishops Assail Gay Marriages as a Threat," *New York Times*, March 10, 2004.

150. "The two talked about homosexuality . . .": For a fuller description of their early work (from which my brief description is drawn, as well as from my interviews with Jeannine Gramick), see Robert Nugent and Jeannine Gramick, *Building Bridges: Gay and Lesbian Reality and the Catholic Church* (Twenty-Third

Publications, Mystic, CT, 1992); "exposing the limitations . . .": ibid., p. 189; "full range of Church teachings . . .": Jeannine Gramick and Robert Nugent, *Response to the Maida Commission Report*, "Magisterial Teaching on Homogenital Behavior," no. A1, January 12, 1995 (published by *National Catholic Reporter—NCR—*online); see *www.natcath.com/NCR_Online/documents/gramnuge.htm* for documents quoted in this chapter; "intrinsically disordered . . .": "Letter to the Bishops of the Catholic Church on the Pastoral Care of Homosexual Persons," Congregation for the Doctrine of the Faith, October 1, 1986, no. 3; "moral neutrality of a homosexual orientation . . .": *Response to Maida Commission Report* (see also *Building Bridges* for Church's distinction between a homosexual orientation and homosexual acts, p. 139); "immorality of prejudice and discrimination . . .": ibid.

150. "Seminars in more than 130 . . .": Gramick and Nugent, *Building Bridges*, p. 206; "Always Our Children . . .": *A Pastoral Message to Parents of Homosexual Children and Suggestions for Pastoral Ministers*, Bishops' Committee on Marriage and Family (USCCB, Washington, D.C., 1997).

151. "surveyed 20 dioceses and 52 parishes . . .": *Special Report: In Search of Best Practices in Ministry with Gay and Lesbian Catholics*, (Center for Applied Research in the Apostolate, Georgetown University, Washington, D.C.), Summer 2004; "Many in the Church suspect . . .": Gramick and Nugent, *Building Bridges*, p. 104. "It is a time in which the faith community . . .": ibid.

152. "Separate themselves totally . . .": Letter from Apostolic Pro-Nuncio, Pio Laghi, to Most Reverend Adam J. Amida, appointing Maida to head committee to evaluate Gramick's and Nugent's work, May 9, 1988 (*NCR* online); "harshly negative tone . . .": "Response to the Report of the Findings of the Commission Studying the Writings and Ministry of Sister Jeannine Gramick," no. 1 (*NCR* online); "engaged in an important and needed ministry . . .": "Report of the Findings of the Commission Studying the Writings & Ministry of Sister Jeannine Gramick, SSND, and Father Robert Nugent, SDS," October 4, 1994, p. 3 (*NCR* online); "Considerable testimony among evangelical Protestant communities . . .": ibid., p. 10.

153. 'deplorable' that homosexuals have been the object of violent . . .": "Letter to the Bishops of the Catholic Church on the Pastoral Care of Homosexual Persons"; "condone . . . surprised when . . .": ibid.; "erroneous and dangerous positions . . .": *Contestatio*—"Erroneous and Dangerous Propositions in the Publications *Building Bridges* and *Voices of Hope* by Sister Jeannine Gramick, SSND, and Father Robert Nugent, SDS," Congregation for the Doctrine of the Faith, October

24, 1997, p. 2 (*NCR* online); "Erroneous conscience . . .": Response of Sister Jean-nine Gramick, SSND, to the Congregation for the Doctrine of the Faith's *Contesta-tio*, "Introduction," Item c, February 5, 1998 (*NCR* online). "objective moral evil . . .": ibid., no. 1b, p. 4; "to present the full range of the Church's teachings . . .": ibid., Con-clusion.

154. "because I never disclosed . . .": Jeannine Gramick, "Response to the Con-gregation for the Doctrine of the Faith," July 29, 1998 (*NCR* online); "homosexual acts are always . . .": Congregation for the Doctrine of the Faith, "Profession of Faith," undated (*NCR* online); "submission of will and intellect . . .": ibid.; "errors and ambiguities . . .": Congregation for the Doctrine of the Faith, *Notification Regard-ing Sister Jeannine Gramick, SSND and Father Robert Nugent, SDS* (USCCB, May 31, 1999); "Downplayed one aspect of Church teaching . . .": Lisa Sowle Cahill, "Silenc-ing of Nugent, Gramick sets a novel standard of orthodoxy," *America*, Vol. 181, Issue 4, August 14–21, 1999; "the Vatican policy on conformity . . .": ibid.; "non-infallible teaching . . .": ibid.; "bottom line of orthodoxy . . .": ibid.

155. "more beneficial to minister . . .": "Statement of Jeannine Gramick, SSND, Regarding Discernment On the Notification of the Congregation for the Doctrine of the Faith," September 23, 1999, pp. 2–3 (*NCR* online); "reconsidered and, hopefully, ultimately reversed . . .": ibid.; "the negative effects of the CDF deci-sion . . .": ibid.; "I choose not to collaborate . . .": Statement from Sister Jeannine Gramick, in response to May 23, 2000 silencing order, May 25, 2000 (*NCR* online).

158. "But in 2003, in its statement denouncing . . .": *Considerations Regarding Proposals to Give Legal Recognition to Unions Between Homosexual Persons*, June 3, 2003, no. 4; "Who embrace and celebrate . . .": "Our Foundational Letter to the Church," Pentecost Sunday 1998, *www.rainbowsash.com*.

159. "61 percent of Catholics believe . . .": Cathy Lynn Grossman, "Poll: Catholics view fundamentalists less favorably," *USA Today*, November 16, 2001, refers to Zogby International/Le Moyne College nationwide phone survey of 1,508 Roman Catholics; "69 percent thought homosexual acts . . .": Jerry Filteau, "Survey finds Catholics skeptical of bishops handling of sex abuse," *Catholic News Service*, July 21, 2004, refers to study by sociologists Dean R. Hoge, Catholic University, and James D. Davidson, Purdue University, sponsored by University of Notre Dame, of 1,119 adult Catholics; "more favorable opinion of gay men and lesbians . . .": "Reli-gious Beliefs Underpin Opposition to Homosexuality," Pew Research Center for the People and the Press, November 18, 2003, p. 2 (national survey of 1,515 adults, including Catholics); "An estimated half of Catholics would allow . . .": CBS News

Poll, "U.S. Catholics Angry at Church," May 2, 2002; "(54 percent) of white Catholics . . .": "Religious Beliefs Underpin Opposition to Homosexuality," p. 4; "Condone . . . the homosexual lifestyle . . .": Archbishop Harry Flynn, "Archbishop Flynn explains reason for rescinding catechesis award," *Catholic Spirit*, May 29, 2003.

160. "One year later . . .": Martha Sawyer Allen, "Gays make quiet showing at mass," *Star Tribune*, April 24, 2004.

161. "To experience the charity of Christ . . .": "Pastoral Care of Divorced and Remarried," Address of Pope John Paul II to the Pontifical Council for the Family, prepared for the Internet by Msgr. Peter Nguyen Van Tai, Radio Veritas Asia, Philippines, January 24, 1997; "obstinately persist in manifest . . .": "Declaration by the Pontifical Council for Legislative Texts," Julian Herranz, titular Archbishop of Vertara, President (Vatican City, June 24, 2000); "must always decline . . .": "Address of John Paul II to the Prelate Auditors, Officials and Advocates of the Tribunal of the Roman Rota," January 28, 2002, no. 9.

162. "Angel Meacham, a teacher . . .": Liz Oakes, "Catholic teachers mostly disagree with Kentucky firing," *Cincinnati Enquirer* online edition, September 27, 2003; "diocesan officials received a letter . . .": Kevin Eigelbach, "Church says it won't pry," *Kentucky Post* online edition, September 25, 2003; "Vicki Manno sued . . .": AP story, Marie McCain contributor, "Cleveland Catholic school sued over firing," *Cincinnati Enquirer* online edition, September 28, 2003; "from burning dinner to adultery": Christine Schenk, "Jesus and Women" brochure (FutureChurch); "seen in this light . . .": ibid.

163. "338 in 1968 . . .": Kenneth C. Jones, *Index of Leading Catholic Indicators: The Church Since Vatican II*, (Oriens Publishing, 2003); "47,000 in 2002 . . .": *Annuarium Statistucim Ecclesaie*, 2002 Statistical Yearbook of the Church, Vatican Secretariat of State; "Almost 70 percent of the nearly 70,000 . . .": ibid.; "40 percent of all marriages . . .": Charles N. Davis, "Divorce and Remarriage: The Sense of the Faithful," *Rome Has Spoken*, p. 166; "80 percent never apply . . .": George M. Anderson, "Marriage Annulments: An Interview with Father Ladislas Orsy," *America*, Vol. 177, no. 9, October 4, 1997; "Rauch, an Episcopalian . . .": Sheila Rauch, *Shattered Faith* (Henry Holt and Company, New York, 1997).

165. "At the 1980 Synod of Bishops . . .": "Divorce and Remarriage," p. 167, Davis quoting Philip Kaufman, OSB, *Why You Can Disagree and Remain a Faithful Catholic* (Crossroad, NY, 1995), p. 128;

166. "nearly three-quarters of Catholics disagree . . .": ibid., p. 166, Davis referring to 1992 Gallup survey; "Under 21 percent . . .": William V. D'Antonio, et al,

American Catholics: Gender, Generation an[...]
78; "an estimated 80 percent to 90 perce[...]
nulments in America," www.catholic.n[...]
.html.

167. "When she arrived at a marriage[...]
traditio/august.asp; "Divorce is a grave offen[...]
(Doubleday, NY, 1994), nos. 2384, 2385.

168. "of heroic love . . .": Pope John Paul I[...]
Given Us!" Homily for the Beautification of l[...]
Mora and Gianna Beretta Molla (Vatican, May 1[...] ...any marital
difficulties . . .": ibid.; "compromised the serenity [...] ∪sservatore Romano, April
24. 1994, as published in "Biographies of New Blesseds," *www.ewtn.com*; "to the
physical and psychological violence . . .": ibid.

169. "Abstinence-until-marriage sex education . . .": "Fact Sheet: Abstinence
Only Sex Education," July 2004, *www.plannedparenthood/org/library/facts/Abstinence
Only10.01.html*; "A Brief Explanation of Federal Abstinence-Only-Until-Marriage
Funding," Sexuality Information and Education Council of the United States,
www.siecus.org/policy/states/BriefExplanationofFederalAb-OnlyFunding.pdf; "World
Health Organization considers programs . . .": A. Grunseit and S. Kippax, "Effects
of Sex Education on Young People's Sexual Behavior" (Geneva: World Health Or-
ganization, 1993), as quoted by Debra W. Haffner, "Abstinence-only Sexuality Education Programs Try to Scare Young People Chaste, but Let's Get Real," *Conscience*,
Spring, 1998; "more than half of all high-school students . . .": "The Facts: School
Condom Availability," Advocates for Youth, *www.advocatesforyouth.org/publications/
factsheet/FSSCHCON.HTM*.

170. "research shows . . . use condoms . . .": ibid.

9. Catholic Women for a New Sexual Ethic

172. "We are still teaching a sexual code . . .": Gudorf, *Body, Sex, and Pleasure*,
pp. 2–3; "bodyright . . .": ibid., p. 161; "must be done on a text by text basis . . .": ibid.,
p. 11; "procreationism . . . focuses moral authority . . .": ibid., p. 24; "moral minimalism . . .": ibid., p. 15.

173. "If a licit sexual act . . .": ibid.; "unconscionable that many forms . . .": ibid.,
p. 17; "the unwritten law embedded . . .": ibid., p. 62; "reversing the prevailing understanding . . .": ibid., p. 46; "The failure to examine embodied . . .": ibid., p. 65; "no
function save": ibid.

174. "If the place[...] ibid., p. 139; "the[...] Church's T[...] Church[...]

...ment of the clitoris . . .": ibid.; "accepting mutual pleasure . . .": ...dangers of lust . . .": Sydney Callahan, "Stunted Cathechesis: The ...ching on sexuality has a role in its current crisis," in *C21 Resources-The* ...*in the Twenty-first Century*, Boston College newsletter, Fall 2003, p. 5.; "When ...e psychosexual": ibid.; "values of a wholesome sexuality . . .": *Human Sexuality: New Directions in American Catholic Thought*, A Study Commissioned by the Catholic Theological Society of America (Paulist Press, 1977), p. 95.

175. "self-liberating . . .": ibid., pp. 92–95; "personally frustrating . . .": ibid., p. 95; "Such a revisionist view": Rosemary Radford Ruether, "Sex and the Body in Catholic Tradition," *Conscience*, January 31, 2000, p. 9; "it stands as an important expression . . .": ibid.; "life-cycle ceremonies . . .": Rosemary Radford Ruether, *Christianity and the Making of the Modern Family* (Beacon, Boston, 2000), p. 214; "the more diverse forms of family . . .": ibid.

176. "sexual friendship covenants . . .": ibid., p. 215; "a few friends and mentors together": ibid.; "veiled in lies . . .": ibid.; "to learn about their own sexuality . . .": Rosemary Radford Ruether, "Sexual Illiteracy," *Conscience*, summer 2003, p. 17; "permanently committed relationship . . .": *Christianity and the Making of the Modern Family*, ibid.; "lifelong effort . . .": ibid.; "remain faithful . . .": ibid., p. 216; "every covenanting relationship . . .": ibid., p. 217; "often find themselves abandoned . . .": ibid.; "partners who are dissolving . . .": ibid.

177. "once the church . . ." ibid., p. 214; "in the goodness of sexual embodiment . . .": Callahan, "Stunted Catechesis," p. 5; "When the Catholic tradition . . .": "Let's Talk about Sex," an interview with Fran Feder, *U.S. Catholic*, February 2003; "For woman" (actually, reads "For Man . . .") *Gaudium et Spes*, Chapter 1, "Dignity of the Human Person, no. 16.

10. New Foundations: Catholic Women Reshape Theology

178. "The Catholic community . . .": Rosemary Radford Ruether, "For feminist theologians, a good job is hard to find," *National Catholic Reporter*, January 15, 1999; "Women represent nearly one-third . . .": "Head Count Enrollment by Degree Program, Race or Ethnic Group and Gender," Fall 2003, U.S. Schools, Data Table 2.13, from The Association of Theological Schools in the United States and Canada.

179. "In the last 25 years . . .": "How many women are enrolled in theological schools?" *Fact Book on Theological Education—2002–2003*, The Association of Theological Schools in the United States and Canada, p. 18; "All of the numeric gain . . .": ibid.; "majority of students—55 percent . . .": "Head Count Enrollment," Data Table

2.13; "26 percent of the nearly 5,500 . . .": ibid.; "At the nineteen theological schools . . .": 2004 Theological School Faculty Survey, Center for the Study of Theological Education, Auburn Theological Seminary, New York, NY; "fully 78 percent of today's women . . .": Mary Ann Hinsdale, IHM, "Jubilee: New Horizons in Theological Education," *New Horizons in Theology* (Orbis, 2005); "In 1970, only 2 percent . . .": Barbara G. Wheeler, "True and False," First in a series of reports on theological school faculties, *Auburn Studies*, No. 4, January 1996; "the critical principle . . .": Rosemary Radford Ruether, *Sexism and God-Talk, with a New Introduction* (Beacon Press, Boston, 1993), pp. 18–19; "has created the very conditions . . .": Catherine Mowry LaCugna, "Catholic Women as Ministers and Theologians," *America*, October 10, 1992, p. 248.

180. "most feminist theologians . . .": ibid., 247; "the fashioner of all things": Wisdom 7:22; "listening, welcoming . . .": "Letter to the Bishops of the Catholic Church on the Collaboration of Men and Women;" "God did not just speak . . .": *Sexism and God-Talk*, p. xiv; "On the contrary . . .": ibid.

181. "what Johnson was *really* thinking . . .": I am indebted to Mary Gordon for my introduction to Elizabeth Johnson and her experience before the American cardinals; see Mary Gordon, "Women of God," *Atlantic Monthly*, January 2002; "The Women Stayed . . .": Artist Sister Helen David Brancato, 1993.

182. "a small group of elite officeholders or canonized saints . . .": Elizabeth A. Johnson, "Friends of God and Prophets: Toward Inclusive Community," speech at Call to Action conference, Milwaukee, Wisconsin, November 4, 2000; "A Message of Hope and Courage . . .": *Madeleva Manifesto*, St. Mary's College, Notre Dame, Indiana, April 29, 2000; "Carry forward the cause . . .": ibid.; "We deplore, and hold ourselves . . .": ibid.

183. "Can a Male Savior Save Women?": *Sexism and God-Talk*, pp. 116–138.

185. "Dysfunctional patriarchal household . . .": For more on this, see Elizabeth A. Johnson, *Truly Our Sister: A Theology of Mary in the Communion of Saints* (Continuum, New York/London, 2003), p. 33.

188. "one whose very nature . . .": Elizabeth Johnson, *She Who Is: The Mystery of God in Feminist Theological Discourse* (Crossroad Publishing, New York, 1992, 2002 Tenth Anniversary Edition), p. 243; "I am who I am": Exodus 3:14; "Linguistically this is possible . . .": *She Who Is*, p. 243; "If the mystery of God . . .": ibid.

190. "There are serious doubts . . .": "Study, Prayer Urged Regarding Women's Ordination 'Responsum'—CTSA Board Received Committee Paper," *Origins*, Vol. 27, No. 5, June 19, 1997, p. 75.

191. "not all traditions are 'legitimate' . . .": ibid., p. 78; "doctrinally and theologically unthinkable . . . :" National Conference of Catholic Bishops, "Response to CTSA Report on Women and Ordination," *Origins*, Vol. 27, No. 16, October 2, 1997. "a theological wasteland . . .": Cardinal Bernard Law, "The CTSA: A Theological Wasteland," June 18, 1997, *www.dotm.org/wasteland-law.htm.*

192. "public dissent": Carmel McEnroy, *Guests in Their Own House: The Women of Vatican II* (Crossroad, New York, 1996), pp. 273–279; "seriously deficient in [her] duty . . .": ibid.; "accusations from unnamed . . .": Pamela Schaeffer, "Seminary ousts Fiand after push from right," *National Catholic Reporter*, May 1, 1998; "only remaining female full professor . . .": ibid.

193. "Forced exile . . .": Leslie Wirpsa, "Before silencing, Gebara speaks her mind," *National Catholic Reporter*, August 25, 1995; "Out of the Depths . . .": Ivone Gebara, *Out of the Depths: Women's Experience of Evil and Salvation*, translated by Ann Patrick Ware (Fortress Press, Minneapolis, 2002).

194. "No matter how impressive . . .": Statement from the National Coalition of American Nuns, "Seminary Women: Out in the Cold," May 12, 1998.

196. "The voices of women . . .": Susan A. Ross, "Teaching feminist theology to college students: The influence of Rosemary Radford Ruether," *Cross Currents*, Spring 2003.

197. "Mary, Mary quite contrary . . .": Elizabeth Johnson, Excerpt from *Truly Our Sister*, *U.S. Catholic*, December 2003; "truly sister to our strivings . . .": *Truly Our Sister*, p. xvii; "The doctrinal mythologizing approach. . . .": ibid., p. 40.

198. "What begins as praise . . .": ibid., p. 266; "The Magnificat is buried . . .": Ruth Fox, "Women in the Bible and the Lectionary," *Liturgy 90*, May/June, 1996. "As the women's movement . . .": Mary Daly, *Beyond God the Father* (Beacon Press, Boston, 1973), p. 13–14.

199. "Over the long run . . .": Peter Steinfels, *A People Adrift: The Crisis of the Roman Catholic Church in America* (Simon & Schuster, 2004), p. 277.

11. Catholic Women Raise the Dead

202. "That action 'must have been understood . . .'": Elisabeth Schüssler Fiorenza, *In Memory of Her: A Feminist Theological Reconstruction of Christian Origins* (Crossroad, New York, 2002, Tenth Anniversary Edition), p. xliv; "are engraved . . .": ibid., p. xliii; "Nag Hammadi Library . . .": "About the Nag Hammadi Library Section," "Dead Sea Scrolls Collection/The Nag Hammadi Library," "An Introduction to Gnosticism and the Nag Hammadi Library," The Gnostic Society Library,

www.gnosis.org; "rivaled the apostle Peter's . . .": "The Gospel According to Mary Magdalene," *www.gnosis.org/library/marygosp.htm.*

203. "Junius . . .": Barbara Kantrowitz and Anne Underwood, "The Bible's Lost Stories," *Newsweek,* December 8, 2003, p. 52.

204. "in what is still . . .": Dorothy Irvin, "Pilgrims see how early Christian women celebrated the Eucharist," *Calendar 2003: The Archaeology of Women's Traditional Ministries in the Church 100 to 820 A.D.* (U.S., 2002), January-February; "Here lies . . .": Irvin, *Calendar 2004: The Archaeology of Women's Traditional Ministries in the Church 300–1500 A.D.* (U.S., 2003), March-April; "scarcely a ten-minute walk . . .": "Bishop Theodora in Women's Episcopal Succession," *Calendar 2003,* September-October; "indicate that she celebrates . . .": "Vitalia, A Priest Celebrates Mass," *Calendar 2004,* July-August; "The pictures shown . . .": ibid., Introduction.

205. "While 'the widely held assumption . . . '": Fox, "Women in the Bible and the Lectionary"; "passage is omitted from Matthew's . . .": ibid.; "John's account . . .": ibid; "anointing of Jesus by the woman . . .": ibid.; "very significant omission . . .": ibid.; "Catholics never hear Romans." ibid., and Regina A. Boisclair, "Amnesia in the Catholic Sunday Lectionary: Women—Silenced from the Memories of Salvation History," in *Women and Theology,* Mary Ann Hinsdale (Orbis, 1995), p. 119.

206. "on no Sunday . . . women in Acts . . .": ibid.; "In fact . . . most Catholics . . .": Christine Schenk, *Women in the Lectionary: Suggestions for Celebrating An Inclusive God in Jesus Christ* (Future Church, undated); "As for Mary Magdalene": ibid.; "Inexplicably . . . this important account . . .": ibid.

207. "Detroit, 2002, some 150 women and men . . .": Description of the Mary Magdalene service based on my viewing of a videotape produced by Call to Action and interviews with Marge Orlando, Eileen Burns, Therese Terns, and Karon Van Antwerp.

12. The New Face of Catholic Ministry

212. "I have felt a call . . .": Alexandra Guliano, *Leadership for Transition: The Parish Director and the Lay Pastored Parish in the Roman Catholic Church,* unpublished dissertation, San Francisco Theological Seminary, April 21, 2000, p. 104; "emphatically told that how I experienced God . . .": ibid., p. 105.

217. "In the late 1990s women held . . .": *Women in Diocesan Leadership Positions: A Progress Report,* edited version of report prepared by William Daly, National Association of Church Personnel Administrators (USCCB, Washington, D.C., 2002), and Anne Munley, Rosemary Smith, Helen Maher Garvey, Lois MacGillivray, Mary

Milligan, *Women and Jurisdiction: An Unfolding Reality/The Leadership Conference of Women Religious Study of Selected Church Leadership Roles* (LCWR, USA, 2001); "In 1965, there were 58,632 priests . . .": "Frequently Requested Church Statistics," Center for Applied Research in the Apostolate (CARA), Georgetown University, Washington, D.C., 2004 (http://cara.georgetown.edu).

218. "more priests in United States over age of ninety . . .": "Priest Shortage: Statistics and Trends," Ministry Formation Directory (CARA, 2003); "The United States has a ratio of one priest to every 1,400 . . .": "Frequently Requested Church Statistics"; "In Latin America . . . in Africa . . .": Mary L. Gautier and Brian Froehle, *Global Catholicism: Portrait of a World Church* (Orbis Books, 2003); "more than 20 percent of all Catholic seminarians . . .": Mary L. Gautier, *Catholic Ministry Formation Enrollments: Statistical Overview for 2003–2004*, (CARA, March 2004); "of the country's roughly 19,000 parishes . . .": "Frequently Requested Church Statistics"; "In Dubuque, Iowa": "Parishes without Priests," *Religion and Ethics Newsweekly*, PBS, episode no. 642, June 20, 2003; "Projections are that in the years ahead . . .": Moderator Tim Russert, referring to writing of Sister Mary Johnson, at launch of "The Church in the Twenty-first Century" initiative, published in "From This Church Forward," *Boston College Magazine*, Fall 2003; "Worldwide, at least one-quarter . . .": "Frequently Requested Church Statistics"; "nearly thirty thousand Catholics . . .": Philip J. Murnion and David DeLambo, *Parishes and Parish Ministries: A Study of Parish Lay Ministry* (National Pastoral Life Center for the National Conference of Catholic Bishops, New York, 1999), p. iii; "nearly 180,000 nuns . . .": "Frequently Requested Church Statistics."

219. "some two hundred parishioners per parish . . .": Murnion, *Parishes and Parish Ministries*, p. 3; "A study of 473 Catholic campus ministries . . .": Mark M. Gray, PhD and Mary E. Bendyna, RSM, PhD, "Catholic Campus Ministry: A Report of Findings from CARA's Catholic Campus Ministry Inventory.," (CARA, September 2003), p. 57; "almost twenty-six thousand lay Catholics . . .": Gautier, *Catholic Ministry Formation Enrollments*; "Two-third of those students are women . . .": ibid.; "just 3,200 graduate-level seminarians . . .": ibid.

223. "The priest is not just a functionary . . .": Archbishop James Kelcher and Bishops Stanley Schlarman, George Fitzsimmons, and Eugene Gerber. "Pastoral Message of Kansas Bishops on Sunday Communion without Mass," June 18, 1995, *www.ewtn.com*; "scarcity of priests . . .": Pope John Paul II, Encyclical Letter *Ecclesia De Eucharistia* to Bishops, Priests, Deacons, Men, and Women in the Consecrated Life, and Lay Faithful on the Eucharist and its Relationship to

the Church, April 17, 2003, no. 33; "In some places . . . the practice of the Eucharistic . . .": ibid., no. 10; "profound grief . . .": ibid.; "The Eucharist is too great a gift . . .": ibid.

224. "In the early 2000s, the Vatican, and . . . U.S. Conference . . .": "Extraordinary Ministers of Holy Communion at Mass—General Principles," Committee on Liturgy, USCCB, May 21, 2002, and "General Instruction of the Roman Missal," USCCB, June 3, 2003; "special permission from Rome . . .": "New Norms Set for the Distribution and Reception of Holy Communion under Both Bread and Wine," USCCB, June 3, 2003; "In 2003, the Vatican was working . . . liturgical abuses . . .": "Outdated liturgy document creates a stir," *National Catholic Reporter*, October 3, 2003.

225. "Regardless of how much time . . .": Guliano, *Leadership for Transition*, p. 18.

230. "Number of ordained deacons . . .": "Frequently Requested Church Statistics."

231. "Ordained women deacons until the fifth century . . .": Phyllis Zagano, "Women's Ministry in the Catholic Church," *Hofstra Horizons*, Fall 2003, and Phyllis Zagano, *Holy Saturday: An Argument for the Restoration of the Female Diaconate in the Catholic Church* (Crossroad Publishing, New York, 2000); "Why not take the woman . . .": Ann Rodgers-Melnick, "Catholics debate purpose of women deacons," *Pittsburgh Post-Gazette*, April 21, 2002.

232. "the eighth sacrament . . .": Archbishop Sean P. O'Malley, *Chrism Mass Homily*, April 6, 2004; "when necessity demands it . . .": Vatican Congregation for Divine Worship and the Discipline of the Sacraments, in collaboration with the Congregation for the Doctrine of the Faith, *Redemptionis Sacramentum: On certain matters to be observed or to be avoided regarding the Most Holy Eucharist*," March 25, 2004, no. 161.

235. "Everyone knows that . . .": Luke Timothy Johnson, "Sex, Women and the Church: The Need for Prophetic Change," *Commonweal*, 103:2, June 20, 2003.

236. "strong advocates for greater lay involvement . . .": Ginny Ramsey, Catholic Action Center co-director, as quoted by Frank E. Lockwood, "Catholic bishop fires five officials," *Lexington Herald-Leader*, April 7, 2004; "affirmation of faith . . .: Aviva L. Brandt, "Bishop Requires Lay People to Sign Oath," Associated Press," July 20, 2004.

237. "In 2003, for the first time in nearly 20 years . . .": Gautier, *Catholic Ministry Formation Enrollments*; "Two-thirds of the students . . .": ibid.; "wholeheartedly toward incorporating . . .": Sr. Katrina Schuth, *Seminaries, Theologates, and the Future of*

Church Ministry: An Analysis of Trends and Transitions(Liturgical Press, Collegeville, Minnesota, 1999), p. 130; "According to Schuth, an estimated one-third of Catholic seminaries . . .": ibid., p. 56.

238. "What I see in the broadening of this definition of sacrament . . .": Susan A. Ross, *Extravagant Affections: A Feminist Sacramental Theology* (Continuum, New York/London, 2001), p. 213.

13. One Woman who Refused to Wait: The Ordination of Mary Ramerman

241. "In his book about his seminary days . . .": James Brady Callan, *The Studentbaker Corporation: A Vehicle for Renewal in the Catholic Church* (Spiritus Publications, Rochester, NY, 2001), Other resources for this chapter were Ramerman's and Callan's workshop presentations at Call to Action Conference, September 14, 2003.

243. "others were overjoyed . . .": ibid., p. 71; "Soon Ramerman was participating . . .": ibid.; "Mary couldn't get away . . ." ibid., p. 73.

244. "Besides Ramerman . . .": ibid., pp. 70, 73.

245. "a 'trustworthy priest' in his place . . .": ibid., p. 78; "You can't hold back spring . . .": ibid., p. 81; "Statement of Faith . . .": ibid., p. 84, see also "The Story of Spiritus Christi,"*www.spirituschristi.org/history/html*; "If the Church stands up . . .": ibid., p. 88.

246. "Never again allow a woman . . .": ibid., p. 93.

247. "These six people . . .": ibid., p. 96.

248. "had chosen to excommunicate . . .": Rob Cullivan, "Spiritus Christi 'solidifies' schism," *Catholic Courier Online*, Vol. 113, No. 6, November 8, 2001.

249. "Ludmila Javorova . . .": Miriam Therese Winter, *Out of the Depths: The Story of Ludmila Javorova, Ordained Roman Catholic Priest* (Crossroad Publishing Company, New York, 2001).

250. "Gibson reportedly is building his own church . . .": Christopher Nixon, "Is the Pope Catholic . . . Enough?" *New York Times Magazine*, March 9, 2003.

251. "one cannot honor . . .": Cullivan, "Spiritus Christi 'solidifies' schism"; "in a teaching position . . .": ibid.; "while Gibson may be a bit . . .": John L. Allen, Jr., "Pope gives 'The Passion' thumbs-up," *National Catholic Reporter*, December 26, 2003.

14. Women's Ordination: Calling the Question

258. "'wounded' the Church": Joseph Cardinal Ratzinger and Tarcisio Bertone, "Decree of Excommunication," Sacred Congregation for the Doctrine of the Faith, August 5, 2002.

259. "During my studies": Ida Raming testimony, in *Called to be a Woman Priest*, edited by Ida Raming, Gertrud Jansen, Iris Muller and Merchtilde Neuendorff, translation by Mary Dittrich (Austria, 1998); "prerequisite for justice for women": ibid.; *"We Shall Keep Quiet No More!"*: Published in Zurich, 1964; "Published in 1976 . . . :" Ida Raming, *The Exclusion of Women from the Priesthood: Divine Law or Sex Discrimination*, translated by Norman R. Adams, (Scarecrow Press, Metuchen, New Jersey, 1976).

261. "James Keady . . .": "Ex-Student Sues St. John's Over Contract for Nike Gear," *New York Times*, November 29, 1999, and Mark Conrad, "Ex-Student, Coach Sues St. John's University, Nike," *www.sportslawnews.com/arcive/articles%201999/St John'ssuit.htm*.

264. "inviting Raming to preside . . .": Beth Waltemath, "The Elements of Excommunication," *The Turning House*, Student Journal of Union Theological Seminary, Vol. 3, No. 3, May 12, 2003.

266. "For today, the continuing discrimination . . .": Ida Raming, "Women's Ordination—God's Gift for a Renewed Church," speech given during five-week U.S. tour, April-May 2003; "Now they speak . . .": ibid.; "It is time that this web . . .": ibid.; "prophetic sign of protest . . .": ibid.

269. "Reverend Joanna Adams was assigned . . .": David Van Biema, Elisabeth Kauffman, Jeanne McDowell, Marguerite Michals, Frank Sikora and Deirdre Van Dyk, "Rising Above the Stained-Glass Ceiling, *Time*, June 28, 2004.

THE CHURCH AND REFORM: LIST OF PROGRESSIVE ORGANIZATIONS

✝——✝——✝

Association for the Rights of Catholics in the Church (ARCC)

Mission: To bring about substantive structural change in the Catholic Church, including shared decision-making and accountability.

Founded in 1980.

P.O. Box 85

Southampton, MA 01073

Telephone: 413-527-9929

Fax: 413-527-5877

www.arcc-catholic-rights.net

BishopAccountability.org

Mission: To document the sex abuse crisis and hold Catholic bishops accountable by assembling an Internet collection of every publicly available document and report.

Founded in 2003.

P.O. Box 541375

Waltham, MA 02454-1375

Telephone: 781-910-5467

www.bishop-accountability.org

Call to Action USA

Mission: To work for equality and justice in the Church and society.

Founded in 1976.

2135 W. Roscoe 1N

Chicago, IL 60618
Telephone: 773-404-0004
Fax: 773-404-1610
www.cta-usa.org

Catholic Network for Women's Equality (CNWE)

Mission: To enable women to name their giftedness and to effect structural change in the Church, reflecting the mutuality and coresponsibility of women and men.
Originally founded in 1981; established as CNWE in 1988.
P.O. Box 84524
2336 Bloor Street, Toronto, ON
M6S 4Z7
Canada
www.cnwe.org

Catholic Organizations for Renewal (COR)

Mission: A coalition of Catholic organizations working together for reform and renewal of the Catholic Church, and for justice and peace.
Founded in 1991.
18 Ruggles Street
Melrose, MA 02176
Telephone: 781-665-5657
Fax: 718-665-5066
www.cta-usa.org/COR.html

Catholics for a Free Choice

Mission: To shape and advance sexual/reproductive ethics that are based on justice, reflect a commitment to women's well-being, and affirm women's and men's moral authority.
Founded in 1973.
1436 U Street NW, Suite 301
Washington, D.C. 20009-3997

Telephone: 202-986-6093
Fax: 202-332-7995
www.catholicsforchoice.org

Catholics Speak Out

Mission: To encourage Church reform, particularly equality, justice, and dialogue between laity and hierarchy on issues of sexuality, sexual orientation, and reproduction.
Founded in 1987 (a project of the Quixote Center).
P.O. Box 5206
Hyattsville, MD 20782
Telephone: 301-699-0042
Fax: 301-862-2182
www.quixote.org/cso

CITI (Celibacy Is the Issue) Ministries/Rent-A-Priest

Mission: To locate, recruit, and promote the availability of married and other resigned Roman Catholic priests to fill the spiritual needs of Catholics.
Founded in 1992.
P. O. Box 2850
Framingham, MA 01703-2850
Telephone: 800-774-3789
www.rentapriest.com

Coalition of Catholics and Survivors

Mission: Catholics, survivors, and advocates working for immediate and practical solutions to the sex abuse crisis in the Catholic Church.
Founded in 2002.
396 Washington Street, H294
Wellesley, MA 02481
Telephone: 781-396-7196
www.catholicsandsurvivors.net

Corpus—The National Association for an Inclusive Priesthood

Mission: To promote an expanded and renewed priesthood of married and single men and women in the Catholic Church.
Founded in 1974.
114 Sunset Drive
Raynham, MA 02767-1383
Telephone: 508-822-6710
Fax: 508-822-6710
www.corpus.org

Dignity USA

Mission: To work for respect and justice for gay, lesbian, bisexual, and transgender persons in the Catholic Church and the world through education, advocacy, and support.
Founded in 1969.
1500 Mass. Ave. NW, Suite #8
Washington, D.C. 20005-1894
Telephone: 800-877-8797 or 202-861-0017
Fax: 202-429-9808
www.dignityusa.org

Federation of Christian Ministries

Mission: To empower and prepare people for ministry, including Roman Catholic women, and to provide ministry certification.
Originally founded in 1968; established as Federation of Christian Ministries in 1973.
1905 Bugbee Road
Ionia, MI 48846-9663
Telephone: 800-538-8923
www.federationofchristianministries.org

FutureChurch

Mission: National coalition of parish-based Catholics seeking the full participation of all baptized Catholics in Church life; addresses, e.g., women in ministry, optional celibacy.

Founded in 1990.

15800 Montrose Avenue

Cleveland, Ohio 44111

Telephone: 216-228-0869

Fax: 216-228-4872

www.futurechurch.org

Good Tidings

Mission: To help women and priests in clandestine relationships, women in exploitative relationships with priests, and mothers of priests' children, to make life decisions.

Founded in 1983.

P.O. Box 283

Canadensis, PA 18325-0283

Telephone: 570-595-2705

Fax: 570-595-6096

www.marriedpriests.org/GoodTidings.htm

Healing Voices, Inc.

Mission: Lay Catholics and people of faith who promote healing and advocacy for survivors of sexual abuse by clergy and religious leaders by building a caring community.

Founded in 2001.

P.O. Box 142

Frederick, MD 21705

Telephone: 301-788-2663

www.healingvoices.org

International Movement We Are Church

Mission: A global network of groups working to create dialogue to bring about the renewal and reform of the Roman Catholic Church.
Founded in 1996.
www.we-are-church.org

Las Hermanas

Mission: Hispanic women united to empower themselves and others to participate in the transformation of Church and society by sharing Hispanic culture, language, spirituality, and traditions.
Founded in 1971.
E-mail: hermanasUSA@sbcglobal.net

Leadership Conference of Women Religious

Mission: Association of leaders of Catholic congregations of women religious in the United States; concerns have included transforming women's place in Church and society.
Founded in 1956.
8808 Cameron Street
Silver Spring, MD 20910-4113
Telephone: 301-588-4955
Fax: 301-587-4575
www.lcwr.org

The Linkup

Mission: To foster healing, prevention, and education in the area of clergy sexual abuse.
Founded in 1991.
P.O. Box 429
Pewee Valley, KY 40056
Telephone: 502-241-5544
Fax: 502-241-0031
www.thelinkup.org

Mary's Pence

Mission: A Catholic organization committed to women's empowerment that collects and distributes funds to Catholic women's grassroots ministries throughout the Americas.

Founded in 1987.

402 Main Street, Suite 210

Metuchen, NJ 08840

Telephone: 732-452-9611

Fax: 732-452-9612

www.maryspence.org

National Coalition of American Nuns

Mission: To study and speak out on issues of justice in Church and society.

National Office

12434 Klinger Street

Detroit, MI 48212

Telephone: 313-891-2192

Network

Mission: National Catholic social justice lobby that educates, lobbies, and organizes to influence the formation of federal legislation to promote economic and social justice.

Founded in 1971.

801 Pennsylvania Avenue SE, Suite 460

Washington, D.C. 20003-2167

Telephone: 202-547-5556

Fax: 202-547-5510

www.networklobby.org

New Ways Ministry

Mission: To provide a gay-positive ministry of advocacy and justice for lesbian and gay Catholics and reconciliation within the larger Christian and civil communities.

Founded in 1977.
4012 Twenty-ninth Street
Mt. Rainier, MD 20712
Telephone: 301-277-5674
Fax: 301-864-8954
http://mysite.verizon.net/~vze43yrc

North American Conference of Separated and Divorced Catholics

Mission: Healing and recovery of those who have experienced separation and divorce.
Founded in 1974.
P.O. Box 10
Hancock, MI 49930
Telephone: 906-482-0494
Fax: 906-482-7470
www.nacsdc.org

Partners in Preaching

Mission: To prepare and support women and men, ordained and lay, to share in the Church's ministry of liturgical preaching.
Founded in 1997.
7136 Arbor Glen Drive
Eden Prairie, MN 55346-3114
Telephone: 952-975-9470
www.partnersinpreaching.org

Quixote Center

Mission: Justice and peace organization working for structural change in the Church and society; operates various programs, including Priests for Equality and Haiti Reborn.
Founded in 1976.
P.O. Box 5206
Hyattsville, MD 20782
Telephone: 301-699-0042

Fax: 301-862-2182

www.quixote.org

Save Our Sacrament (S.O.S.)/Reform of Annulment and Respondent Support Network

Mission: A support and reform network for Catholic respondents in annulment proceedings, with a focus on their rights.

Founded in 1997.

P.O. Box 5119

Cochituate, MA 01778

www.saveoursacrament.org

Soulforce

Mission: An interfaith movement committed to ending spiritual violence perpetuated by religious policies and teachings against gay, lesbian, bisexual, and transgender people.

Founded in 1998.

P.O. Box 3195

Lynchburg, VA 24503-0195

Telephone: 877-705-6393

Fax: 434-384-9333

www.soulforce.org

Survivors First

Mission: To ensure that sexual abuse survivors are aware of their recovery options, and to keep children safe; emphasis on collaboration among survivors and nonsurvivors.

Founded in 2002.

P.O. Box 81-172

Wellesley Hills, MA 02481

Telephone: 781-910-5467

www.survivorsfirst.org

Survivors Network of those Abused by Priests

Mission: Volunteer self-help organization that seeks healing for survivors and for the institution that hurt them; offers support groups, education, and prevention.

Founded in 1989.

P.O. Box 6416

Chicago, IL 60680-6416

Telephone: 1-877-SNAPHEALS

www.snapnetwork.org

Voice of the Faithful

Mission: To provide a prayerful voice, attentive to the Spirit, through which the faithful can actively participate in the governance and guidance of the Catholic Church; supports those abused by priests and priests of integrity, and shapes structural change in the Church.

Founded in 2002.

P.O. Box 423

Newton Upper Falls, MA 02464

Telephone: 617-558-5252

Fax: 617-558-0034

www.votf.org

Women-Church Convergence (WCC)

Mission: A coalition of autonomous, Catholic-rooted organizations and groups raising a feminist voice and working for the empowerment of women in Church and society.

Originally founded in 1977 (Women of the Church Coalition); became WCC in 1984.

c/o Dr. Bridget Mary Meehan

5856 Glen Forest Drive

Falls Church, VA 22041

Telephone/Fax: 703-671-1972

www.women-churchconvergence.org

WomenPriests.org

Mission: Largest international website on the ordination of women, created by Roman Catholic theologians.

Founded in 1999.

www.womenpriests.org

Women's Alliance for Theology, Ethics, and Ritual (WATER)

Mission: An international community of justice-seeking people who promote the use of feminist values to make religious and social change.

Founded in 1983.

8035 Thirteenth Street

Silver Spring, MD 20910

Telephone: 301-589-2509

Fax: 301-589-3150

www.his.com/mhunt

Women's Ordination Conference

Mission: Works locally, nationally, and internationally (in collaboration with the worldwide movement) for the ordination of women as priests and bishops in a renewed priestly ministry in the Roman Catholic Church.

Incorporated in 1977.

P.O. Box 2693

Fairfax, VA 22031-0693

Telephone: 703-352-1006

Fax: 703-352-5181

www.womensordination.org

Women's Ordination Worldwide (WOW)

Mission: To promote worldwide Roman Catholic women's ordination to a renewed priestly ministry in a democratic Church and stand in solidarity with women who are ordained in the ongoing renewal of the Church.

Founded in 1996 (umbrella group of national and international organizations; leader/location changes every two years).

www.womensordination.org/pages/intern_wow.html; see also www.wow 2005.org

Other Resources:

"The Archaeology of Women's Traditional Ministries in the Church," calendar series (2003, 2004, 2005) and notecards, available only from *irvincalendars@hotmail.com*

THE CHURCH AND REFORM:
TIMELINE OF
SELECTED EVENTS

✠ ⎯✠⎯ ✠

1960s   Second Vatican Council meets. (1962–65)

Pope Paul VI issues *Humanae Vitae*, maintains ban on artificial birth control. (1968)

Graduate theology programs in the United States and Europe begin to admit women.

1970s   Ludmila Javorova ordained a priest by Bishop Felix Maria Davidek in underground Catholic Church in former Czechoslovakia. (1970)

The first women—the "Philadelphia 11"—ordained Episcopal priests. (1974)

Women's Ordination Conference holds first national conference. (1975)

U.S. bishops hold Call to Action conference (1976), out of which forms the leading national progressive reform group, Call to Action.

Pontifical Biblical Commission concludes there are insufficient scriptural grounds to rule out women's ordination. (1976)

Vatican issues *Inter Insigniores*, declaring women cannot be priests because they do not image Christ. (1977)

Women's Ordination Conference holds second national conference. (1978)

Karol Wojtyla becomes Pope John Paul II. (1978)

Sister Theresa Kane appeals publicly to Pope John Paul II to open "all ministries of our Church" to women. (1979)

1980s: Center of Concern mounts "Women Moving Church" conference. (1981)

First meeting of independent women-church groups (1983), followed by formation of Women-Church Convergence.

Catholics for a Free Choice runs *New York Times* ad that a diversity of opinion exists on abortion. (1984)

Sister Jeannine Gramick and Father Robert Nugent are ordered by Church officials to separate themselves from New Ways, their pioneering ministry to Catholic homosexuals. (1984)

Vatican publishes *Letter to the Bishops of the Catholic Church on the Pastoral Care of Homosexual Persons*, which condemned violence against homosexuals, but also said that no one "should be surprised" that violence occurs when homosexual activity is condoned. (1986)

Pope John Paul II issues *Mulieris Dignitatem, On the Dignity and Vocation of Women*, defining women's and men's roles as different and complementary, in the Church and in the world. (1988)

U.S. bishops attempt to write a Pastoral Letter on Women (1983 to 1992), without success.

1990s John Paul II issues *Ordinatio Sacerdotalis, On Reserving Priestly Ordination to Men Alone*, declaring case closed on women's ordination. (1994)

Vatican Congregation for the Doctrine of the Faith releases *Responsum ad Dubium, Concerning the Teaching Contained in Ordinatio Sacerdotalis*, declaring the ban on women priests to be infallible teaching. (1995)

Pope John Paul II releases his "Letter to Women" in preparation for the United Nation's Fourth World Conference on Women in Beijing, denouncing discrimination against women, but again stressing women's complementary roles. (2003)

International Movement We Are Church presents a petition with 2.5 million signatures to the Vatican supporting a range of progressive Church reforms. (1997)

Catholic Theological Society of America takes a stand against Vatican assertion that ban on women priests is infallible teaching. (1997)

U.S. bishops publish pastoral on homosexuals, *Always Our Children: A Pastoral Message to Parents of Homosexual Children and Suggestions for Pastoral Ministers*. (1997)

Vatican issues *Ad Tuendam Fidem (for the Defense of Faith)*, extending canon law to excommunication for those who reject "definitively held" teachings, including the ban on women priests. (1998)

Sister Jeannine Gramick and Father Robert Nugent publicly ordered by Vatican to stop all ministry to Catholic homosexuals. (1999)

The Holy See obstructs consensus on reproductive health issues at major UN conferences on women, including the International Conference on Population and Development in Cairo (1994) and Fourth World Conference on Women in Beijing. (1995)

Catholics for a Free Choice launches the SeeChange Campaign to altar the Holy See's UN status as a Non-member State Permanent Observer, a status unique among the world's religions. (1999)

2000s Sister Jeannine Gramick refuses silencing order, issued by her religious community, on all speech about homosexuals and the Church. (2000)

Madeleva Manifesto, "A Message of Hope and Courage," is published, signed by leading feminist theologians. (2000)

FutureChurch begins in 2000 to promote special celebrations on the feast of Mary of Magdala, July 22, which grow to thousands of people attending hundreds of celebrations worldwide.

National Catholic Reporter breaks story of the abuse of nuns by priests worldwide, based on reports Catholic sisters had been turning over to the Vatican since the 1990s. (2001)

With the support of the Erie Benedictines, Sister Joan Chittister rejects Vatican silencing order and speaks at the first world conference of Women's Ordination Worldwide. (2001)

Mary Ramerman defies Catholic Church authorities and is ordained a Catholic priest. (2001)

Borrowing from the sanctity of life campaign, Catholics for a Free Choice launches its "Condoms4Life" campaign on World AIDS Day, 2001.

The *Boston Globe* breaks stories of priests' crimes against minors and institutional cover-up, inspiring an avalanche of similar stories nationwide. (2002)

U.S. bishops open their Dallas meeting on clergy child sex abuse to the press and public, publish a Charter on the Protection of Children and Young People, set up Office of Child and Youth Protection, and appoint an independent National Lay Review Board. (2002)

Seven women (the Danube 7) ordained Catholic priests on the Danube River; a month later, the Vatican excommunicates them. (2002)

More than four thousand attend first national conference of new moderate reform group, Voice of the Faithful. (2002)

Under pressure from survivors, laity, priests, and the press, Boston Cardinal Bernard Law resigns in disgrace. (2002)

"Religion and the Feminist Movement" conference mounted by Harvard Divinity School's Women's Studies in Religion Program. (2002)

Membership in the Survivors Network of those Abused by Priests climbs; survivors and lay Catholics join forces for the first time in new organizations to work for change. (2002–2003)

Danube 7 priest Ida Raming tours twelve U.S. cities in five weeks. (2003)

Voice of the Faithful holds New York regional conference. (2003)

Vatican issues *Lexicon on Ambiguous and Colloquial Terms about Family Life and Ethical Questions*, with a chapter denouncing Catholics for a Free Choice. (2003)

Vatican issues *Considerations Regarding Proposals to Give Legal Recognition to Unions Between Homosexual Persons*, condemning homosexual unions as immoral and insisting that governments must contain the phenomenon. (2003)

Vatican publishes *Doctrinal Note on Some Questions Regarding the Participation of Catholics in Political Life*, demanding a defense of the basic right to life from conception to natural death, including the rights of the human embryo. (2003)

U.S. bishops publish *Faithful Citizenship: A Catholic Call to Political Responsibility*, which instructs Catholics to advocate for the full range of Catholic social teachings. (2003)

U.S. bishops release results of their audit of diocesan compliance with the terms of the Charter for the Protection of Children and Young People. (2003)

One hundred sixty-three Milwaukee priests sign a letter to U.S. Conference of Catholic Bishops' president urging the opening of the priesthood to married men; similar actions follow by other priests' groups and reform groups like Call to Action and FutureChurch. (2003–2004)

U. S. bishops publish John Jay College survey of incidence of child sexual abuse among Catholic clergy and National Lay Review Board survey on causes and context for the crisis. (2004)

Barbara Rick's documentary *In Good Conscience: Sister Jeannine Gramick's Journey of Faith* opens to sold-out crowds in New York City. (2004)

An uproar of alarm greets the pope's statement reiterating his view of women: *On the Collaboration of Men and Women in the Church and in the World.* (2004)

After some U.S. bishops threaten to withhold Communion from pro-choice politicians and those who vote for them, the full bishops' conference determines that the decision should be left to individual bishops. (2004)

Boston College sponsors "Envisioning the Church Women Want" conference, part of its Church in the Twenty-first Century series. (2004)

The Vatican appoints Cardinal Law to head a major basilica in Rome. (2004)

The Farm, the first national treatment center for clergy sex abuse survivors, opens in Louisville, Kentucky, founded by a survivors' group, The Linkup. (2004)

Pope John Paul II dies; Cardinal Joseph Ratzinger becomes Pope Benedict XVI. (2005)

ACKNOWLEDGMENTS

A book like this requires the participation of so many people that I am afraid to name any for fear of failing to name them all. Certainly, I must begin by thanking every woman who trusted me enough to share her experience with me, both those named in the book and those not named but who are here in spirit. I must especially thank Sister Joan Chittister, who opened her life and her home to me early on, a gesture of trust and kindness that cast a bright light for me on the road ahead. For that, and for the privilege of knowing her, I am deeply grateful.

I thank all of my dear friends and colleagues who gave me their unending support, who believed there would be a book long before I believed it, including Claude Barilleaux, Donna Seaman, Rose Brown, Rochelle Green, Cindy Cooper, Mother Susanna Williams, Jmel Wilson, Jean Brewer, Marie Wilson, Sara Gould, Helen LaKelly Hunt, Kanyere Eaton, Joanne Edgar, Mary Thom, Amy Pouser-Webb, Becky Cherian, Marilyn Scott, Judith Hendra, and Nancy Eisenman.

I thank my editor Cassie Jones for her encouragement, commitment, and invaluable help in shaping this book, and my dear friend, agent, and a talented editor, too, Andrea Pedolsky, who helped midwife this book from an idea to its birth. I thank Joanna Parson, who transcribed many hours of tape, and Erin Hanley and Juhie Bhatia for their research assistance. I thank the people at the professional organizations and research centers who were a great help to me, including Mary Gautier at the Center for Applied Research in the Apostolate, Nancy Merrill at the Association of Theological Schools, Dee Christie at the Catholic Theological Society of America, and Sharon Miller at the Center for the Study of Theological Education at Auburn Theological Seminary, with a very special thanks to Boston College theology professor Sister Mary Ann Hinsdale.

I thank my family, including Jim and Robbie Bonavoglia and Fran Mayer, for seeing to it that I didn't lose my sense of humor. And I extend my profound gratitude to my dear mother, Frances Bianco, a deeply devoted Catholic woman whose faith is personal and profoundly felt, who sees God with a poet's heart and an artist's eye, who taught me to find my own way to God and is my model for a faith-filled life.

Finally, to my beloved husband, Glenn Mayer, goes my deepest thanks. I thank him for his unconditional love, his patience and care, for feeding me when I was hungry, making me laugh when nothing seemed funny, urging me to "just write" when I wanted to give up, and for everywhere and all the time, giving me a place—the best place in all the world—to rest my head. To Glenn, I am eternally grateful.

FURTHER READING

Farley, Margaret A. *Compassionate Respect: A Feminist Approach to Medical Ethics and Other Questions*. New York/Mahwah, NJ: Paulist Press, 2002.

Fiorenza, Elisabeth Schüssler. *Wisdom Ways: Introducing Feminist Biblical Interpretation*. Maryknoll, NY: Orbis Books, 2001.

Fox, Thomas C. *Sexuality and Catholicism*. New York: George Braziller, 1995.

Fox, Zeni. *New Ecclesial Ministry: Lay Professionals Serving the Church*. Chicago: Sheed and Ward, 2002.

Gebara, Ivone. *Longing for Running Water: Ecofeminism and Liberation*. Translated by David Molineaux. Minneapolis: Fortress Press, 1999.

Hunt, Helen LaKelly. *Faith and Feminism—A Holy Alliance*. New York: Atria Books, 2004.

Isasi-Díaz, Ada María. *Mujerista Theology: A Theology for the Twenty-First Century*. Maryknoll, NY: Orbis Books, 1996.

Johnson, Elizabeth A., ed. *The Church Women Want: Catholic Women in Dialogue*. New York: Crossroad, 2002.

Jung, Patricia Beattie, Mary E. Hunt, Radhika Balakrishnan, eds. *Good Sex: Feminist Perspectives from the World's Religions*. New Brunswick, NJ: Rutgers University Press, 2002.

Lee, Bernard J. *The Catholic Experience of Small Christian Communities*. New York/Mahwah, NJ: Paulist Press, 2000.

Maguire, Daniel C. *Sacred Choices: The Right to Contraception and Abortion in Ten World Religions*. Minneapolis: Fortress Press, 2001.

McClory, Robert. *Faithful Dissenters: Stories of Men and Women Who Loved and Changed the Church*. Maryknoll, NY: Orbis Books, 2000.

Schneiders, Sandra M. *With Oil in Their Lamps: Faith, Feminism, and the Future*. New York/Mahwah, NJ: Paulist Press, 2000.

Torjesen, Karen Jo. *When Women Were Priests: Women's Leadership in the Early Church and the Scandal of their Subordination in the Rise of Christianity*. San Francisco: HarperSanFrancisco, 1995.

Weaver, Mary Jo. *New Catholic Women: A Contemporary Challenge to Traditional Religious Authority.* Bloomington: Indiana University Press, 1995.

Wijngaards, John. *The Ordination of Women in the Catholic Church: Unmasking a Cuckoo's Egg Tradition.* New York: Continuum, 2001.

Wills, Garry. *Papal Sin: Structures of Deceit.* New York: Doubleday, 2000.

INDEX

Cardinal Bernard Law
Scalia
Clarence Thomas ?
John Geoghan